Who Really Makes
Environmental Policy?

Edited by Sara R. Rinfret

Who Really Makes Environmental Policy?

*Creating and Implementing
Environmental Rules and Regulations*

TEMPLE UNIVERSITY PRESS
Philadelphia • Rome • Tokyo

TEMPLE UNIVERSITY PRESS
Philadelphia, Pennsylvania 19122
tupress.temple.edu

Library of Congress Cataloging-in-Publication Data

Names: Rinfret, Sara R., editor.
Title: Who really makes environmental policy? : creating and implementing
    environmental rules and regulations / edited by Sara R. Rinfret.
Description: Philadelphia : Temple University Press, 2021. | Includes
    bibliographical references and index. | Summary: "This edited volume
    provides students with an explanation of federal and state rulemaking
    processes and regulatory policy and why this context is important
    specifically for U.S. environmental policy. It also includes
    illustrative case studies in each chapter that will allow students to
    apply theory to practice"—Provided by publisher.
Identifiers: LCCN 2021003980 (print) | LCCN 2021003981 (ebook) | ISBN
    9781439920183 (cloth) | ISBN 9781439920190 (paperback) | ISBN
    9781439920206 (pdf)
Subjects: LCSH: Environmental policy—United States. | Environmental
    policy—United States—Decision making. | Environmental policy—United
    States—Case studies. | Environmental law—United States.
Classification: LCC GE180 .W47 2021 (print) | LCC GE180 (ebook) | DDC
    363.7/05610973—dc23
LC record available at https://lccn.loc.gov/2021003980
LC ebook record available at https://lccn.loc.gov/2021003981

9 8 7 6 5 4 3 2 1

Contents

Preface

The origins of this book date back to 2006 in Flagstaff, Arizona, sitting in Dr. Jacqueline Vaughn's Environmental Regulation course at Northern Arizona University. Because of this course, I began to understand why U.S. environmental policy remained in gridlock and the pathways around it. Specifically, the vast majority of environmental policy is not made in the halls of Congress. Instead, environmental policy is created by agency experts (rule writers) in federal environmental agencies (e.g., the Environmental Protection Agency, Fish and Wildlife Service, National Park Service, Bureau of Land Management) and implemented at the state level. These individuals have been delegated the authority to interpret vague legislation and write rules. These rules carry the same weight as congressional law.

Although there are a number of American government, public policy, and environmental policy textbooks on the market, few discuss the role of agency rulemaking and regulatory processes, and the ones that do are outdated. Twenty-first-century undergraduate and graduate students interested in environmental policy courses are not just political science students; they come from a range of disciplines, including environmental studies, history, forestry, public health, public administration, and engineering. These students are future rule writers, environmental inspectors, policy analysts, and state actors who will be responsible for carrying out U.S. environmental policy.

What is missing in classrooms today is a clear understanding of regulatory policy, rulemaking processes, and their centrality in U.S. environmental policy making. The goal of this edited volume is to provide (1) an explanation of rulemaking processes and regulatory policy—why this context is important for U.S. environmental policy—and (2) real-world examples in each chapter help students understand how theory is applied in practice. This edited volume, a first of its kind, is collectively written by experts in U.S. environmental rulemaking and regulatory policy. The chapters are rooted in data and explanatory case studies.

Acknowledgments

First, I want to acknowledge my coauthors in this edited collection. After countless conversations over the years, we are excited for this book to finally come to fruition. Collectively, we share a commitment and passion to ensure that our students truly understand who really makes environmental policy. Regulation matters, and we devote these pages to recognizing and honoring the individuals across federal, state, and local entities that do the hard work.

Additionally, I want to thank my 2020 PUAD 595: Environmental Policy and Governance class for their forward and positive thinking. In the middle of the COVID-19 pandemic, they continuously saw the bright spots in how a moment in time can become the catalyst for change. This change is definitely not easy, but these students pushed me to complete this book during a worldwide pandemic. They embraced my never-ending excitement to share the world of environmental regulation with them, and for this, I am deeply indebted. Our class theme song will always be Warren G's "Regulate."

This book would not have been possible without the meticulous editing by Katie Stilp and Mel Brittner Wells. Katie Stilp, a former colleague from Green Bay, Wisconsin, spent countless hours reading drafts of this manuscript. Mel Brittner Wells's careful reading of draft chapters made several of them even stronger.

I also want to recognize our editor, Aaron Javsicas, who provided much help along the way. We cannot forget the entire Temple University Press staff, who have been a wonderful team, and I thank them all. We would also like to thank the anonymous reviewers who offered their feedback throughout the writing of this book.

Indeed, I have tremendous gratitude toward all the individuals who supported this endeavor. Any and all mistakes are entirely my own.

Sara Rinfret, March 2021
Missoula, Montana

Who Really Makes
Environmental Policy?

1

Setting the Stage

Environmental Rulemaking and Regulation Defined

SARA R. RINFRET

A cross the United States, thousands of biologists, lawyers, economists, geologists, anthropologists, engineers, and sociologists, to name a few disciplines, hold positions in federal, state, and local agencies to implement environmental policy. The vast majority of these individuals are career civil servants responsible for overseeing food safety and safe drinking water or designating critical habitat to prevent the polar bear in Alaska from going extinct. We rarely read about these individuals on our daily Twitter newsfeed. Instead, social media sites (e.g., Twitter, Facebook, Instagram) engender doom scrolling about U.S. environmental policy because of a deeply divided Congress.

However, these negative perspectives, we argue, bring opportunity for alternative understanding. Environmental policy in the twenty-first century is often not made in the halls of Congress. In this edited collection, we argue that understanding the nuances of U.S. environmental regulations is imperative. Our focus is simple: We want readers to understand how federal environmental agencies (rule writers) and state inspectors implement environmental policy. For everyday Americans, environmental policy is complex and, for some, falls short of expectations. After reading this text, readers should understand how environmental policy is quietly guided by experts in regulatory agencies to weather a bumpy political storm.

This chapter serves as a foundation, where we introduce the theme of the text—understanding the evolution of environmental rulemaking and regulations from creation to implementation. Subsequent chapters are guided by rich case studies or real-world examples to foster among students an understanding of the connection of theory to practice and an appreciation of the pathways and opportunities for environmental policy outside of Congress.[1] Collectively, the chapters serve as the baseline for an environmental policy tomorrow, the descriptive framework described in Chapter 8.

The focus in this introductory chapter is to examine the process and implementation of U.S. environmental regulations. Specifically, I describe why and how regulations are made and what happens after a regulation goes into effect. This chapter challenges us to rethink the notion that environmental policy is made only in Congress. Instead, environmental policy is shaped and defined across federal and state agencies.

Context Defined

Many college courses do not spend much time defining regulation or the actors involved. We believe it is important to provide context into the interconnectedness between the so-called fourth branch of government (bureaucracy), environmental policy, regulation (and its process of rulemaking), and its unsung heroes (front-line actors). As we know, the United States has three major branches of government—legislative, executive, and judicial. Within this system of governance, each branch provides checks and balances on the others.

For example, when Congress (the legislative branch) passes a law, the president (the executive branch) must sign for the law to go into effect. Unfortunately, we do not spend enough time discussing with students what occurs after a congressional law is enacted. Congress delegates its lawmaking authority to administrative agencies (the bureaucracy), where policy is implemented. An administrative agency is also known as a federal organization or agency (e.g., the Environmental Protection Agency, U.S. Fish and Wildlife Service, Bureau of Land Management, National Park Service). Bureaucrats are the individuals such as scientists and policy analysts who work in these agencies. We begin with a discussion of why Congress delegates its lawmaking authority to agencies before turning to defining regulations, its process (rulemaking), and an overview of the chapters in this text.

Congressional Delegation

When Congress grants policy-making power to federal agencies, it is described as delegation of authority. There are numerous motivations for why delegation of authority occurs, and they warrant explanation. We identify three areas—setting guidelines, shifting responsibility, and expertise—to explain why Congress delegates its lawmaking authority to agencies.

Setting clear and specific guidelines. The first explanation for delegation of authority is because members of Congress do not want to specify particular details about programs in legislation. Doing so could be detrimental for their reelection efforts.[2] We use a coal-burning power plant to illustrate this point. Let's consider Congress creates legislation setting guidelines for allowable emission levels (a cap) for coal-burning power plants across the United States. These new standards make the air cleaner, reducing health consequences. However, such parameters could also lead to layoffs in a congressional member's district if the coal company does not have the financial resources to install new technology to meet targets. In turn, citizens of these districts might become angry, harming the Congress member's reelection efforts. Alternatively, members of Congress leave legislation ambiguous so that experts in an administrative agency will use their expertise to interpret and implement the law.

Shifting the responsibility. Vague congressional legislation informs the second rationale for delegation of authority—shifting responsibility.[3] Members of Congress shift their policy-making authority to agencies to shift the blame to agencies. We use another example—critical habitat designation for the polar bear—to explore this rationale. Instead of Congress passing legislation setting aside a specific habitat area to protect the polar bear from becoming extinct, the U.S. Fish and Wildlife Service (USFWS) can research the implications of various scenarios. For example, a decision about the designation of habitat decision could anger oil and gas companies because a habitat area is within a large oil

TABLE 1.1 CHECKS AND BALANCES	
Branch of government	Forms of accountability for bureaucracy
Legislative (Congress)	• Administrative Procedure Act of 1946 • Budgets • Legislation
Executive (the president)	• Presidential appointments • Office of Management and Budget's Office of Regulatory Affairs
Judicial (the courts)	• Litigation

reserve. Leaving policy implementation to the USFWS in this case shifts blame to the experts, not Congress.

Expertise. The final rationale for congressional delegation of authority is expertise. Members of Congress are not experts in each and every policy area. Instead, federal and state agencies are composed of experts in specific policy areas with the knowledge to carry out public policy. For example, a congressional member might not have the expertise to determine the acceptable levels of carbon monoxide for indoor air quality. Experts in the Environmental Protection Agency (EPA) and the Occupational Safety and Health Administration (OSHA) possess the training and educational background to make these determinations.

Delegation of authority may cause us to wonder how unelected bureaucrats can use their expertise to implement public policy. Congress, the president, and the courts serve as accountability checks, and Congress explicitly delegates its policy-making authority. Table 1.1 illustrates how each branch of government serves as a check on agency decision making. We explore the role of each branch in turn.

Congress created the Administrative Procedures Act (APA) of 1946 as the primary oversight mechanism of agency decision making. Agency decision making is also known as rulemaking (discussed later in this chapter). The APA requires agencies to follow its administrative rulemaking guidelines so that government entities can carry out congressional statutes through the creation of rules. An agency rule carries the same weight as a congressional law. Congress requires agencies to provide notice to the public so Americans can participate in processes.

Congressional budgets and legislation serve as additional accountability measures. Congress controls agency budgets. The U.S. Senate Appropriations Committee determines how much money an agency receives for specific programs. If members of Congress are troubled by the policy direction of an agency, the agency's budget can be decreased. Moreover, Congress can write legislation to overturn an agency rule. This approach rarely occurs, because it takes a great deal of effort to create and pass congressional legislation.[4]

The president and Supreme Court play oversight roles in agency decision making. Presidents have the power to appoint (with Senate confirmation) the head of an agency. President Richard Nixon recommended William Ruckelshaus to serve as the first-ever administrator for the EPA. This person serves at the highest level of a federal agency and can determine the policy direction, which often aligns with the president's political perspective. Ruckelshaus, however, demonstrated otherwise. His approach was to ensure the implementation of the EPA's mission "to protect environmental and human health."[5] The president also uses the Office of Management and Budget's (OMB) Office of Regulatory Affairs (OIRA) review. OIRA's role is to review the costs and benefits associated with agency regulations that may be economically burdensome for business (over $100 million). As stated in President Ronald Reagan's Executive Order 12291, the OMB director is authorized to "review any draft proposed or final rule or regulatory impact analysis from a covered agency."[6] Succeeding presidents have continued the practice of OIRA review with the goal of ensuring regulations do not have an undue impact on business.

As Neil Kerwin and Scott Furlong suggest, "No institution of government has been as persistent in its oversight of rulemaking for a longer period of time than the federal judiciary."[7] This statement reminds us that individuals or organizations that are not pleased with the outcome of an agency decision are entitled to seek litigation. For example, the EPA might make a determination to ban chemicals used by dry cleaners. In turn, the dry cleaner's association files suit against the EPA because it does not agree with the rule. The finalization of the rule is pending until courts determine the outcome.

This discussion of the delegation of authority provides insight into the why behind agency decision making and lays the foundation for the next section on defining environmental policy and regulations. Later, the chapter closes with an explanation of agency decision-making processes (rulemaking) and those responsible for ensuring compliance with the law (inspectors).

Environmental Policy, Regulations, and Understanding the Process

Congress deals with several public policy issues on a daily basis, and one of these is environmental policy—government action related to the natural environment.[8] Regulation, by way of comparison, implements legislation. Implementation occurs through a process of rulemaking and compliance (inspections). Environmental regulation is legislation carried out by agencies to protect the natural environment.

The focus of this book is an examination of how environmental agencies at the federal and state level carry out congressional laws. Table 1.2 provides a brief list of federal and state environmental agencies to illustrate the types of organizations tasked with carrying out environmental regulations in the United States. Each federal and state agency contains a mission statement that indicates the expectations of policy implementation by Congress (federal level) or a state legislature.

Administrative rulemaking is the process through which environmental policies are interpreted and defined into rules by rule writers. A rule writer is the overseer of the rulemaking process. His or her training and expertise vary by agency (e.g., environmental science, law, policy, wildlife biology). This is a step-by-step process (see Figure 1.1).

Once a rule is final, to ensure compliance and implementation, state-level agencies use an inspection process (see Chapter 5) and state-level rulemaking (Chapter 6) to monitor activities and behaviors of regulated entities (e.g., businesses). The inspection process is guided by state-level regulators (also known as inspectors).

The Rulemaking Process: Step-by-Step Guide

With the initial foundational concepts defined, we offer a step-by-step outline of the rulemaking process. Each of these stages is explored in later chapters through case studies. This chapter introduces the

TABLE 1.2	FEDERAL AND STATE ENVIRONMENTAL AGENCIES		
Federal agency	**Mission statement**	**State agency**	**Mission statement**
U.S. Environmental Protection Agency	To protect human health and the environment[1]	Montana Department of Environmental Quality	Protect, sustain, and improve a clean and healthful environment to benefit present and future generations[2]
U.S. Fish and Wildlife Service	Conserve, protect, and enhance fish and wildlife, and their habitats for the continuing benefit of the American people[3]	Ohio Environmental Protection Agency	Protect the environment and public health by ensuring compliance with environmental laws and demonstrating leadership in environmental stewardship[4]
National Park Service	Preserves unimpaired the natural and cultural resources and values of the National Park System for the enjoyment, education, and inspiration of this and future generations[5]	Colorado Department of Public Health and Environment	To protect and improve the health of Colorado's people and the quality of its environment[6]
Bureau of Land Management	Sustain the health, diversity, and productivity of public lands for the use and enjoyment of present and future generations[7]	Wisconsin Department of Natural Resources	To protect and enhance our natural environment[8]

1. U.S. Environmental Protection Agency, "Our Mission and What We Do," December 7, 2016, https://19january2017snapshot.epa.gov/aboutepa/our-mission-and-what-we-do_.html.
2. Montana Department of Environmental Quality, "Mission Statement and Guiding Principles," deq.mt.gov/DEQAdmin/about/mission.
3. U.S. Fish and Wildlife Service, "Mission Statement," 2007, resources.ca.gov/wetlands/agencies/usfws.html.
4. Ohio Environmental Protection Agency, "About Us," https://epa.ohio.gov/About.
5. National Park Service, "About Us," November 4, 2019, https://www.nps.gov/aboutus/index.htm.
6. Colorado Department of Public Health and Environment, "Strategic Plan 2016–2019 and Department Implementation Plan FY 2017–18," https://www.colorado.gov/pacific/sites/default/files/OPP_CDPHE-2016-2019-Strategic-Plan-FY2017-18.pdf.
7. U.S. Department of the Interior, Bureau of Land Management, "Our Mission," https://www.blm.gov/about/our-mission.
8. Wisconsin Department of Natural Resources, "Our Mission," https://dnr.wi.gov/about/mission.html.

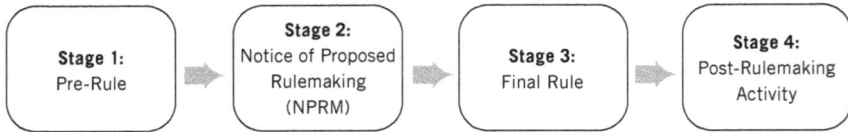

Figure 1.1 Rulemaking Process

basics of the process (see Figure 1.1) and how various institutional entities (Congress, the courts, the president) and noninstitutional actors (stakeholder groups) are involved. The case of biking routes in Mammoth Cave National Park is used to explore the intricacies of rulemaking.[9]

Mountain Biking in Mammoth Cave National Park

Mountain biking is a favorite pastime for many Americans. In 2018 alone, more than eight million Americans participated in mountain biking as a leisure activity. One might assume that mountain biking is an environmentally sustainable activity. However, environmental policy is much more complex than decreasing carbon emissions through sustainable practices such as biking.

Across the United States, national parks confront the impact of mountain biking on native plants and species or cultural sites within park boundaries. Since the 1970s, the common approach has been to ban mountain bike use across all national parks, because the language Congress used to create the National Park Service (NPS) indicated a dual purpose. The NPS was created "to promote and regulate the use of the Federal areas known as national parks, monuments, and reservations . . . by such means and measures as conform to the fundamental purpose to conserve the scenery and the natural and historic objects and the wildlife therein and to provide for the enjoyment of the same in such manner and by such means as will leave them unimpaired for the enjoyment of future generations." Since the agency's inception, NPS management and staff have continuously attempted to reconcile the tensions between preserving and promoting resources.

For example, in 2012, the National Park Service director, Jon Jarvis, concluded that "bikes are a great way to exercise, get healthy and experience the great outdoors"; therefore, decisions about prohibiting or allowing biking should be up to specific park unit superintendents. Parks such as Mammoth Cave National Park (MCNP) decided to engage the public through regulatory processes to determine whether mountain biking would be a viable option in carrying out the park's specific mission.

MCNP, located in southeastern Kentucky, contains more than four hundred caves and fifty-three thousand acres of forest. The mission of MCNP is "to protect and preserve for the future the limestone caverns and associated karst topography, and the biologically diverse Green River, but also to provide for scientific study, public education, and sustainable recreation use." Part of the regulatory process is to receive input from the public. In this example, park staff heard from environmental organizations, cave divers, and mountain biking companies. These organizations provided input to park staff about mountain biking in the park and helped to shape the park's path forward. This example is useful to trace through the federal rulemaking process.

Stage 1

As noted earlier in this chapter, one of Congress's accountability mechanisms is the Administrative Procedures Act of 1946, which defines the parameters for an agency expert (rule writer) to initiate the creation of a rule. Figure 1.1 illustrates the stages of how an agency expert creates a rule. Stage 1 is best described as the pre-rule phase, when information gathering transpires. The National Park Service—and specifically, for the case explored in this chapter, Mammoth Cave National Park—has the statutory authority to create a rule regarding bicycle use because it must "conserve the scenery and the natural and historic objectives of the wild."

During this stage, agency rule writers informally discuss with affected parties, prior to promulgation of a rule, their goal of examining new bicycle routes in the park. At this point in the process, the agency can gather information and conduct research to best inform the later publication of a proposed rule. In the MCNP case, agency rule writers met with frequent park visitors, scientists (to determine the impact on vegetation), and outdoor recreation groups, to name a few. These conversations inform the agency about potential concerns or areas of research that should be included in a rule. Chapter 1 delves into the nuances of this stage and the information sharing that occurs between agency rule writers and interest groups.

Stage 2

The informal pre-rule interactions lead to stage 2, when an agency publishes a notice of proposed rulemaking in the *Federal Register.* Recall that a federal agency uses statutory intent from congressional legislation to propose a rule. The *Federal Register* is a daily publication of all U.S. federal agency regulations. Federal administrative agencies are required by the APA to publish and provide notice to the public about a rule. Daily updates of the *Federal Register* can be found at www.federalregister.gov.

On May 11, 2012, the NPS provided notice to the public (notice of proposed rulemaking), inviting comment on adding connector trails for individuals to ride their bikes from campgrounds, for example, to other parts of the park. During this stage, the public has a set time frame to comment on a proposed rule. Comments can be submitted electronically or, in some cases, sent by mail. As Sara Rinfret, Michelle Pautz, and Denise Scheberle note, "Congress provided for a public comment period to ensure agencies considered public input, creating a mechanism for citizen participation in agency decision making."[10]

Typically, a federal agency invites public comments for thirty to sixty days. In the case of seeking input about biking in MCNP, the NPS provided ninety days. If an agency believes more time is necessary, the comment period can be reopened or extended. Importantly, anyone can submit a public comment and voice opinions on a given rule. After the comment period commences, agency personnel are required by the APA to review the comments and respond to commenters. After examining the comments, the agency uses this information to determine the language and substance of the final rule. Depending on the volume of public comments and the capacity of agency staff, it can take an agency one to two years to review public comments.[11] In the case of Mammoth Cave National Park's consideration of biking, it took agency staff four months to complete the review.

Mammoth Cave National Park received twenty-one public comments about adding connector bicycling trails. The comments ranged from suggestions to open more access points in the backcountry for biking to creating shared routes with horseback riding. Some applauded the NPS staff for ensuring effective use for bicycling in the park that does not impact wilderness areas or vegetation. The agency responded to concerns and made changes from the proposed rule to the final rule. Because of participant feedback, for example, the agency changed the speed limit in the park to fifteen miles per hour to ensure public health and safety for visitors. Chapter 2 details the nuances of this stage to determine how and why comments make an impact on environmental policy making.

Stage 3
Once an agency completes stages 1 and 2, a final rule is published in the *Federal Register* if affected parties do not pursue litigation. Although final rules are published electronically on the *Federal Register*, each rule can easily be downloaded. The final rule for the Mammoth Cave case study was published on September 12, 2012.

Stage 4
A final stage involves possible post-rulemaking activities. If a person or group wishes to contest the final rule, litigation or the court system can be used to challenge the final actions of an agency. Depending on the outcome of a court case, a rule cannot be finalized until a judicial decision is made. This was not the case in the Mammoth Cave rule, but Chapter 4 discusses how post-rulemaking activities can impact the outcome of a rule.

State Rulemaking

Although much of this text focuses on the federal rulemaking process, it is worth mentioning that states mirror a similar process. State legislatures create state-level policy. To implement these policies, state agencies must follow the State Administrative Procedure Act (SAPA) set by their respective states (see Chapter 6 for more information).

Most state rulemaking processes begin with an internal agency review to determine whether rulemaking is the best or most appropriate action to address a particular problem. At this phase, agency staff reach out to stakeholder groups to receive input on a policy issue. The process unfolds when the state agency submits a formal proposed rulemaking in a state register. After the agency publishes the notice of proposed rulemaking, a public comment period begins. In the final phase, the state agency provides a complete text of the rule to notify the governor, the president of the state senate, the speaker of the state assembly, and the administrative regulations reviews commission (if a state has one) that a rule is being finalized. The final terms of the state rule are published in the state's register.[12]

State rulemaking is of additional importance when there is an absence of federal policy. For instance, due to the lack of federal policy for hydraulic fracturing (fracking), it is up to states to use rulemaking processes to come up with their own policies (see more on this in Chapter 6).

Environmental Inspectors

Once a federal or state rule becomes a law, regulators and members of the regulated community (businesses and other regulated entities) perform ongoing compliance work. Regulators on the front lines—or street-level bureaucrats—are responsible for the success or failure of policy implementation.[13]

With environmental policy, state regulators are responsible for the vast majority of implementation.[14] For instance, many states have oil and gas regulators whose goal is to ensure compliance with laws such as the Clean Air Act or the Safe Drinking Water Act (see Chapter 6). Typically, these individuals work for their state's environmental protection agency. A state regulator, for example, visits oil and gas production facilities throughout her or his state. If, during one of the visits, emission levels exceed federal limits, the state regulator could work with the business operator to address the issue or close the facility until the issue is remedied (typically for repeat offenders).

A federal regulator, by way of comparison, performs oversight for the mining of coal and the safety of workers. The U.S. Department of Labor suggests underground coal mining is one of the most hazardous occupations in the nation due to concerns surrounding proper ventilation. If ventilation is not provided, deadly results can, and have, occurred. For this reason, the federal Mine Safety and Health Administration regulators have the statutory authority under the Federal Coal Mine Health and Safety Act of 1969 to monitor underground mining practices. Regulators from the Mine Safety and Health Administration examine underground mines to make sure proper equipment and ventilation are provided for miners to help prevent respiratory diseases from coal dust.

State and federal regulators cross a variety of policy areas. To ensure public policy is carried out as intended by the law, regulators conduct site visits. The process to ensure compliance with the law is described as an inspection process or site visit, which Chapter 5 explores in depth. The inspection is where some individuals paint a bad picture of regulators, classifying them as "cops" due to their negative interactions with an inspector during a site visit.

Inspectors are often defined by their interactions with facilities during site visits, over time. These interactions are classified as either precision based or intention based. The precision-based approach is strict and by the book; the regulator is a tough enforcer of the law. Intention-based regulators are more flexible in their approach to compliance and use cooperative means to ensure standards are met, working with the regulated community. Given the situation, a regulator may opt to use a combination of precision- and intention-based enforcement mechanisms to seek compliance from the regulated community.[15]

Highlighting the processes and day-to-day approaches used by rule writers and regulators is especially important for understanding where environmental policy is made in the United States. This edited volume, a first of its kind, is collectively written by the experts in U.S. environmental rulemaking and regulatory policy. Bringing together top scholars in one volume to discuss environmental rulemaking from development (rule creation) to implementation (inspection) advances our understanding of how U.S. environmental policy is made. The chapters are rooted in interview data, explanatory case studies, and real-world examples. The goal is to document why understanding regulatory policy matters for environmental policy tomorrow.

Plan of the Book

To enhance understanding of regulatory policy and its centrality in U.S. environmental policy making, this chapter provides an overview of the regulatory process. Chapters 2 through 7 provide an in-depth

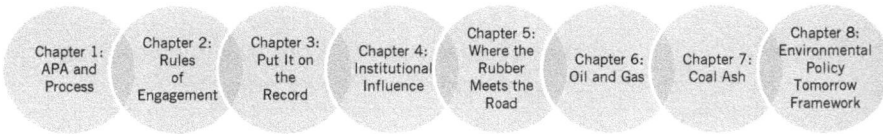

Figure 1.2 Chapter Overview

look into U.S. regulatory policy through an environmental lens. Figure 1.2 offers a brief snapshot of each chapter.

Chapter 2 focuses on the first stage of environmental rulemaking: rule development. The rule development phase is where informal communication occurs between interest groups and agency personnel. At this stage, an agency acquires additional research or discusses a rule with affected entities. Explored are the various rule development approaches used by environmental agencies, from Advanced Notices of Proposed Rulemaking, to regulatory negotiations, to informal conversations with stakeholder groups. The chapter documents how this information serves as a baseline to determine how an agency writes a rule and publishes a notice of proposed rulemaking in the *Federal Register* for public comment. To illuminate how agency rule writers informally gather information to write a rule at this stage, the chapter case explores pet management in the Golden Gate National Recreation Area.

Chapter 3 covers the second stage in the rulemaking process: notice and comment. This stage offers the public the opportunity to influence agency decisions through submitting public comments on proposed rules. This chapter explores and defines the various groups that might participate in a given rule and why they do so. Agency personnel must review and respond to these comments while publicly justifying their preferred course of action. If the justification is inadequate, members of the public can use their comments as the basis to overturn a rule.

Chapter 3 also describes the genesis of the public comment period—to enhance bureaucratic accountability—along with how agencies have sought to engage the public through traditional media, hearings, and, more recently, e-rulemaking. The chapter also summarizes the many pitfalls and challenges associated with relying on notice and comment processes relating to accountability, technical expertise, implementation, and equity. The chapter's case further addresses the challenges of technical expertise and equity by summarizing findings from two climate change–related rulemakings to examine why some members of the public are granted rule changes while others are not.

In the environmental arena, rule finalization can be riddled with controversy. Chapter 4 explores the key institutional (official) actors and their impact on the finalization of U.S. environmental rules. This chapter uses the Endangered Species Act as its case study to provide an in-depth perspective into how Congress, the courts, and the president have attempted to influence twenty-first-century environmental rulemaking. We document the roles of executive orders and the courts and how these impact the role of career civil servants tasked with finalizing an environmental rule. Chapter 4 concludes with a description of what happens after a rule becomes final and stresses the essential role of environmental inspectors and state actors in the implementation—the focus of Chapters 5, 6, and 7.

Crucial to understanding environmental policy on the front lines is investigating how environmental policy is implemented. Chapter 5 tells the story of the implementation of the rules and regulations that are discussed in the earlier chapters. Even after rules are promulgated, implementation is neither straightforward nor easily achieved. Federal environmental rules are delegated to the states for implementation, and state environmental agencies monitor the compliance of regulated firms with these regulations. Inspectors from state—and even sometimes local—government agencies conduct routine inspections to ensure regulatory compliance. The regulatory enforcement approaches or styles of these inspectors are significant in understanding how environmental policy is implemented. Furthermore, the nature of the interactions between the inspector and the regulated community remains a vital, but understudied, area of regulatory compliance. The chapter draws on examples of environmental air quality inspectors to convey the complexities associated with monitoring and enforcing compliance with environmental regulations.

Much like Chapter 5, Chapter 6 explores the role of state and local governments in the implementation of environmental rules. This chapter offers a case study of oil and gas regulation in Colorado. Few issues have been as controversial in Colorado in recent years as oil and gas development. The dramatic increase in drilling activity—much of it in populated areas not accustomed to large-scale energy development—unleashed a political firestorm that shows no signs of abating. This chapter examines how the political furor has shaped (and reshaped) the Colorado Oil and Gas Conservation Commission (COGCC), the state agency responsible for regulating oil and gas development, and oil and gas regulation. Since the early 2000s, the composition of the COGCC has been altered twice, and the state's permitting rules have been amended several times. The state legislature recently fundamentally altered the body's mandate while giving local governments more authority over permitting. The chapter examines how the COGCC has adapted to these changes and how the changes affect permitting and oversight.

Chapter 7 brings to light the interconnectedness between state and federal regulatory actors. This chapter explores how coal, as a longtime fuel source for the generation of electricity, has been in a steady decline recently because of its negative impacts on environmental quality and climate change as well as its inability to compete in the marketplace with cleaner fuels such as natural gas and renewables. Links between the extraction or combustion of coal and pollution problems are well known. However, an increasingly important policy problem examined in Chapter 7 is what has been done to regulate an unwanted by-product: the disposal and storage of coal ash and the risks it poses for water quality and public health. The interplay among multiple policy actors in the regulatory process, including the EPA, state regulators, and utility officials, has resulted in considerable variation across states. Adding to the policy primeval soup is legal uncertainty that surrounds the relationship between coal ash leaks and the subsequent contamination of nearby underground water sources as well as the occasional rise in public visibility and concern that arises from major events in Kentucky and North Carolina.

Partisan Politics and the Importance of Regulatory Processes

In all likelihood, Congress will remain politically divided and subject to policy gridlock for years to come. In 2019, with Republican control of the White House and U.S. Senate, most policies were dead on arrival. This was especially true of environmental initiatives due to a lack

of trust in science to inform decision making and a discourse assert-
ing that regulations are bad for business. With this said, recall that the
first term of the Obama administration maintained a democratically
controlled White House and Congress, health care barely passed, and
environmental policy was not at the forefront. Even with the 2020 election over, the outcome for the presidency and Congress could involve
divisiveness for generations.

Such a condition reinforces the importance of rulemaking and
regulation as an alternative pathway for environmental policy making. In the absence of congressional action, we look increasingly to
agency rule writers and inspectors for critical and innovative approaches to advance U.S. environmental policy. These individuals
are career civil servants who use their expertise to implement environmental policy in an often highly politicized arena. The concluding chapter of this volume highlights our collective work and offers
a guidepost to the environmental policy tomorrow framework. The
elements of this framework (lessons, listening, and leading), we contend, offer direction for a united regulatory journey.

Concluding Thoughts

U.S. environmental policy has undergone significant change due to a
myriad of actors. For decades, the United States led the charge and set
high standards to protect endangered species, clean the nation's waterways, and provide potable water. In 2016, the election of President
Donald J. Trump without a doubt changed the direction of U.S. environmental policy. In his mind, environmental policies, or regulations,
are bad for business, and his administration aggressively attempted to
roll back environmental regulations.[16] Trump rolled back more than
one hundred environmental rules and withdrew the United States
from the Paris Agreement international treaty on climate change. In
his first one hundred days in office, President Biden is already working to undo these rollbacks. The impact of these efforts remains to
be seen.

This does not mean environmental regulations are at a standstill,
but rollbacks are part of defining the historical context of the regulatory apparatus. At the completion of this text, readers will have learned
how environmental policy is made outside of Congress; how regulations

have protected the environment; how public opinion increasingly supports environmental regulations; how business and regulators can and do work together; how states are incubators of environmental innovation; and how agency policy makers (rule writers and inspectors) reconcile external and internal pressures to carry out their agency's missions.

This collective work delves deeply to uncover why environmental policy is not dead in the United States. The U.S. public supports protecting the environment. In 2017, a public opinion poll conducted by Pew Research Center documented that 74 percent of adults believe "the country should do whatever it takes to protect our environment." Because of Americans' willingness to roll up their sleeves and understand the complexities of environmental regulations, there are always positive pathways forward. As we begin this regulatory journey, consider this reminder from Daniel Fiorino: "Regulation has delivered results."[17] As a result, this work serves as the building blocks for environmental policy tomorrow.

Notes

1. Christopher McGrory Klyza and David J. Sousa, *American Environmental Policy: Beyond Gridlock* (Cambridge, MA: MIT Press, 2013).

2. Kenneth Culp Davis, *Discretionary Justice: A Preliminary Inquiry* (Baton Rouge: Louisiana State University Press, 1969).

3. Morris P. Fiorina, "Legislative Choice of Regulatory Forms: Legal Process or Administrative Process," *Public Choice* 39, no. 1 (1982): 33–66.

4. Cornelius Kerwin and Scott Furlong, *Rulemaking: How Government Agencies Write Law and Make Policy* (Washington, DC: CQ Press, 2011).

5. Bill Chappell, "EPA Makes Rollback of Clean Water Rules Official, Repealing 2015 Predictions," NPR, September 12, 2019, https://www.npr.org/2019/09/12/760203456/epa-makes-rollback-of-clean-water-rules-official-repealing-2015-protections.

6. Curtis W. Copeland, "Length of Reviews by the Office of Information and Regulatory Affairs," October 7, 2013, https://www.acus.gov/sites/default/files/documents/Copeland%20Report%20CIRCULATED%20to%20Committees%20on%2010-21-13.pdf, 10.

7. Kerwin and Furlong, *Rulemaking*, 237.

8. Sara Rinfret and Michelle Pautz, *Environmental Policy in Action* (New York: Palgrave, 2019).

9. "Biking Routes in Mammoth Cave National Park," *Federal Register* 77, no. 177 (September 12, 2012), https://www.govinfo.gov/content/pkg/FR-2012-09-12/pdf/2012-22438.pdf.

10. Sara Rinfret, Denise Scheberle, and Michelle Pautz, *Public Policy: A Concise Introduction* (Thousand Oaks, CA: CQ Press, 2018).

11. Kerwin and Furlong, *Rulemaking.*

12. Sara Rinfret, Jeffrey J. Cook, and Michelle Pautz, "Understanding State Rulemaking: Developing Fracking Rules in Colorado, New York, and Ohio," *Review of Policy Research* 21, no. 2 (2014).

13. Michael Lipsky, *Street Level Bureaucrats* (New York: Russell Sage Foundation, 1980); Michael Maynard-Moody and Steven Maynard Musheno, *Cops, Teachers, Counselors: Stories from the Front Lines of Public Service* (Ann Arbor: University of Michigan Press, 2003).

14. Rinfret and Pautz, *Environmental Policy.*

15. Michelle Pautz and Sara Rinfret, *The Lilliputians of Environmental Regulation* (New York: Routledge, 2013).

16. Nadja Popovich, Livia Albeck-Ripka, and Kendra Pierre-Louis, "95 Environmental Rules Being Rolled Back under Trump," *New York Times*, December 21, 2019, https://www.nytimes.com/interactive/2019/climate/trump-environment-rollbacks.html.

17. Daniel Fiorino, *The New Environmental Regulation* (Cambridge, MA: MIT Press, 2006).

Suggested Readings

Kraft, Michael. *Environmental Policy and Politics.* New York: Routledge, 2017.
Scheberle, Denise. *Industrial Disasters and Environmental Policy.* New York: Routledge, 2018.
Vig, Norman, and Michael Kraft. *Environmental Policy: New Directions for the Twenty-First Century.* Washington, DC: CQ Press, 2018.

Bibliography

Chappell, Bill. "EPA Makes Rollback of Clean Water Rules Official, Repealing 2015 Predictions." NPR, September 12, 2019. Available at https://www.npr.org/2019/09/12/760203456/epa-makes-rollback-of-clean-water-rules-official-repealing-2015-protections.
Colorado Department of Public Health and Environment. "Strategic Plan 2016–2019 and Department Implementation Plan FY 2017–18." Available at https://www.colorado.gov/pacific/sites/default/files/OPP_CDPHE-2016-2019-Strategic-Plan-FY2017-18.pdf.
Copeland, Curtis W. "Length of Reviews by the Office of Information and Regulatory Affairs." October 7, 2013, Available at https://www.acus.gov/sites/default/files/documents/Copeland%20Report%20CIRCULATED%20to%20Committees%20on%2010-21-13.pdf.
Culp Davis, Kenneth. *Discretionary Justice: A Preliminary Inquiry.* Baton Rouge: Louisiana State University Press, 1969.

Fiorina, Morris P. "Legislative Choice of Regulatory Forms: Legal Process or Administrative Process." *Public Choice* 39, no. 1 (1982): 33–66.

Gough, Christina. "Number of Participants in Mountain/Non-Paved Surface Bicycling in the United States from 2011 to 2018." *Statista*, March 26, 2019. Available at https://www.statista.com/statistics/763737/mountain-non-paved-surface-bicycling-participants-us/.

Housman, Justin. "Which National Park Properties Allow Bikes on Trails? We Have a List for That." *Adventure Journal*, August 20, 2019. Available at https://www.adventure-journal.com/2019/08/these-are-the-national-park-properties-that-allow-bikes-on-trails/.

Kerwin, Cornelius, and Scott Furlong. *Rulemaking: How Government Agencies Write Law and Make Policy*. Washington, DC: CQ Press, 2011.

McGrory Klyza, Christopher, and David J. Sousa. *American Environmental Policy: Beyond Gridlock*. Cambridge, MA: MIT Press, 2013.

Montana Department of Environmental Quality. "Mission Statement and Guiding Principles." Available at deq.mt.gov/DEQAdmin/about/mission.

National Park Service. "About Us." November 4, 2019. Available at https://www.nps.gov/aboutus/index.htm.

———. "Dueling Mandates: Preservation and Use of National Parks: Yellowstone National Park." Updated August 13, 2015. Available at https://www.nps.gov/teachers/classrooms/dueling-mandate.htm.

———. "Mammoth Cave National Park Business Plan." 2003. Available at https://www.nps.gov/maca/planyourvisit/upload/MACA%20Business%20Plan.PDF.

———. "Mammoth Cave National Park, Kentucky." Updated March 24, 2020. Available at https://www.nps.gov/maca/index.htm.

Ohio Environmental Protection Agency. "About Us." Available at https://epa.ohio.gov/About.

Rinfret, Sara, and Michelle Pautz. *Environmental Policy in Action*. New York: Palgrave, 2019.

U.S. Department of the Interior, Bureau of Land Management. "Our Mission." Available at https://www.blm.gov/about/our-mission.

U.S. Environmental Protection Agency. "Our Mission and What We Do." December 7, 2016. Available at https://19january2017snapshot.epa.gov/aboutepa/our-mission-and-what-we-do_.html.

U.S. Fish and Wildlife Service. "Mission Statement." 2007. Available at resources.ca.gov/wetlands/agencies/usfws.html.

Wisconsin Department of Natural Resources. "Our Mission." Available at https://dnr.wi.gov/about/mission.html.

2

Creating an Environmental Rule

Is It Really Off the Record?

SARA R. RINFRET AND JEFFREY J. COOK

San Francisco, California, is well known for its steep hills, cable cars, and the breathtaking Golden Gate Bridge. Alongside the bridge is the picturesque Marin Headlands, often referred to as the crown jewel of Golden Gate National Recreation Area. Yet, if it were not for the actions of local citizens, Marin Headlands would have become a high-rise condominium complex.

After World War II, San Francisco confronted rapid development and urban sprawl. Much of the remaining open land was owned and operated by the U.S. Army. The army relocated to Southern California, viewing the open space of the Marin Headlands as surplus and selling the land to the highest bidder. As a result, residents expressed concerns about the decreased rural countryside and, in particular, apprehensions surrounding the development of Marin Headlands into a massive condo development. Local citizens organized People for the Golden Gate National Recreation Area, advocating for setting aside this land and protecting it for future enjoyment. Eventually, these efforts led to Congress's creation of Golden Gate National Recreation Area in 1972.[1]

This brief introductory story sets the tone for this chapter. In particular, the checkered and long-standing story of Golden Gate National Recreation Area allow us to uncover the nuances of the first stage of the federal rulemaking process—rule development (often referred to as pre-rule). As noted in Chapter 1, the rule development phase is when informal communication occurs between interest groups or the public and agency personnel. During this stage, agencies acquire additional research, share information, or discuss a rule with affected entities prior to publication of a notice of proposed rulemaking (NPRM).

We begin this chapter with a case study about dog management in Golden Gate National Recreation Area, because it serves as a guidepost for the myriad of actors and details of the rule development phase. Then we explore the various pre-rule engagement mechanisms employed by federal agencies, which include advanced notices of proposed rulemaking (ANPRM), regulatory negotiations, and the role of the Office of Management and Budget (OMB). Importantly, this chapter illuminates how the rule development phase is the essential precursor for an agency to publish an NPRM in the *Federal Register* for public comment (the focus of Chapter 3). This chapter demonstrates the role of stakeholder influence and why this stage in the rulemaking process can take several years.

Case Study: I Can't Take My Dog Where?

Walking one's dog in a national park may seem like a given, but this is not the case for the National Park Service's Golden Gate National Recreation Area (GGNRA). The area contains seventy-five thousand acres, fifty-nine miles of coastline, 1,273 plants and species, and eighty sensitive rare or threatened species.[2] Walking a dog off-leash could negatively affect sensitive ecosystems or pose harm to visitors (e.g., dog bites). However, pet owners in San Francisco might not have access to open space to walk their dogs in the city.

The origins of this case study begin in 1972, when guidelines for pet management were established by the National Park Service (NPS). The NPS policy clearly noted, "Dogs must be on a leash in all areas of national parks."[3] GGNRA, also created in 1972, convened an advisory council about how best to implement this policy. The central question for the council was to determine how to implement the NPS policy for a new park where visitors were not accustomed to restrictions for dog use. Because many locations within the park had allowed off-leash dog walking prior to the creation of GGNRA, the advisory committee determined that pet owners could voice-control their off-leash dogs' behavior.

From the committee's perspective, GGNRA should implement a grandfather clause, allowing specific areas in the park to be designated off-leash voice-control areas. This would assist in the area's transition to a newly designated national park. However, this decision would lead to many years of conflict for park staff and managers.

During the 1990s, visitor use increased with commercial dog-walking companies, and so did conflict between pet owners and non-dog owners. The onus fell on park staff to manage the cleanup of dog waste and the conflict among visitors.[4] In 2001, the NPS determined GGNRA was implementing its policy in error, with increased complaints for dog biting and ecological impact. Although the NPS was silent from 1979 to 2001, GGNRA did not have the authority to enforce its previously adopted voice-control or dog-leash policies. Specifically, the voice-control policy conflicted with the 1972 NPS policy.[5] This burdened GGNRA staff with the herculean task of determining how best to work with the NPS and visitors to manage dog use. In 2001, GGNRA staff embarked on a variety of public input mechanisms before deciding whether to issue a notice of proposed rulemaking.[6]

One of the first public outreach approaches began in 2002. GGNRA staff solicited feedback through an advanced notice of proposed rulemaking. This was a way for park staff to obtain public feedback regarding potential options for pet management. More than eighty-five hundred individuals submitted public comments. Park staff partnered with researchers from Northern Arizona University's Social Science Research Laboratory to contextually examine whether commenters opposed or supported off-leash dog walking. In the researchers' assessment, a strong majority of commenters supported dogs to be on a leash.[7] Senior federal NPS officials reviewed the public comments and suggested that off-leash dog walking would be suitable in select locations of GGNRA. Although this represented the views of many, slightly less than half of the respondents did not support this approach and wanted to ban dogs from the park. As a result, more planning was necessary to move forward to accommodate the desires of the various interests involved.

To gather additional public input, GGRNA established in 2005 a Negotiated Rulemaking (reg neg) Advisory Committee to offer best practices and guidelines for dog management within specified areas of the park. The reg neg committee, according to the Negotiated Rulemaking Act, is limited to a twenty-five-person committee. The goal for GGNRA staff was to work with community members of the potentially affected counties—San Francisco, Marin, and San Mateo.[8] Membership of the committee represented various interests ranging from the NPS, off-leash advocates, dog walkers, ocean beach dog owners, Pacifica dog walkers, Presidio dog walkers, San Francisco dog owners, commercial dog-walking businesses, and environmental organizations such as the Audubon Society, Sierra Club, Marine Mammal Society, and the Center for Biological Diversity.

The intent of the reg neg was for interested parties to work with GGNRA to determine how best to reconcile conflicts surrounding the protection of cultural and natural resources, visitors' safety, and the public desire to walk dogs off-leash in designated areas.[9] The reg neg committee convened a series of public meetings and completed its work in 2007. The committee's recommendations led to the preparation of a draft environmental impact statement to ensure enforceable guidelines for dog walking to prevent harmful ecological impact.[10]

The rule development phase, as described here, took several years, and GGNRA staff's rule development efforts led to an eventual NPRM in 2016. The NPRM focused on designated areas for off-leash dog use.[11] This case study describes the rule development stage, demonstrating the various outreach and rule development processes available to public agencies, which are discussed in this chapter.

The Details: The Characteristics of Rule Development

Chapter 1 defines how the Administrative Procedure Act (APA) requires notice and comment for the public to participate in the rule-making process. However, the APA is silent on how an agency should carry out rule development processes. As a result, agencies have significant flexibility to structure this stage of the process. The rule development process involves three defining iterative phases: agenda setting, fact finding and content development, and review (see Table 2.1).

TABLE 2.1 RULE DEVELOPMENT PHASES	
Agenda setting	President, courts, Congress
Fact-finding and content development	Stakeholder outreach and inclusion mechanisms (advanced notice of proposed rulemaking, negotiated rulemaking committee, informational meetings)
Review	Office of Management and Budget's Office of Information and Regulatory Affairs

Agenda Setting

Institutional actors such as Congress, the courts, or the president shape rulemaking agendas.[12] Congressional statutes structure agency regulatory authority, determining what rules an agency can consider moving forward with. For example, the Environmental Protection Agency has been delegated the authority by Congress to use the Clean Water Act to define what bodies of water are swimmable. Comparatively, Congress created the National Park Service to maintain conservation and recreation for future enjoyment. This is often referred to as a dual mandate; park managers must implement policy (in this chapter's case, dog management) to both preserve the park and ensure enjoyment of the park by its visitors.

Statutory guidance can be unclear, leaving agencies wondering whether they have the discretion to create rules. For example, the EPA was uncertain whether it had the ability to create a rule under the Clean Air Act to regulate greenhouse gas emissions. As a result, a group of Northeastern states sued the EPA (*Massachusetts v. EPA*) for its failure to do so. The court found that the EPA did have the authority to create rules to regulate greenhouse gas emissions under the Clean Air Act, designating the agency the ability to move forward.

The president can also shape rule development. The president uses the Unified Agenda of Federal Regulatory and Deregulatory Actions to document regulatory and deregulatory priorities for agency action.[13] Put simply, the agenda formalizes all the regulatory and deregulatory activities supported by the president across federal agencies.[14] As such, the agenda serves as a vehicle for the president to shape rule development and policy directives for agencies. President Obama stressed the importance of regulating greenhouse gas emissions across federal agencies and tasked the Bureau of Land Management with finding ways to use public lands for wind and solar development. Additionally, Obama believed the regulatory priority for the EPA was to regulate greenhouse gas emissions from cars and coal-burning power plants. President Trump, by way of comparison, signaled his preference for finding mechanisms to eliminate burdensome regulations for business. In the Fall 2019 Unified Agenda, Trump stressed, "By amending and eliminating regulations that are ineffective, duplicative, and obsolete, the Administration can promote economic growth and innovation and protect individual liberty."[15] By way of comparison, the Biden administration wants to modernize the regulatory process. In January 2020, President Biden noted in a memorandum to executive agency department heads, "Our Nation today faces serious challenges, including a massive global pandemic; a major economic downturn; systemic racial inequality; and the undeniable reality and accelerating threat of climate change. It is the policy of my Administration to mobilize the power of the Federal Government to rebuild our Nation and address these and other challenges. As we do so, it is important that we evaluate the processes and principles that govern regulatory review to ensure swift and effective Federal action. Regulations that promote the public interest are vital for tackling national priorities."[16]

Fact Finding and Content Development

The regulatory agenda is a signal to agencies that they can begin fact-finding approaches. Put simply, fact-finding is when an agency collects information about a rule. Information gathering encourages agency personnel to collect material pertaining to the environmental, economic, and public health factors of a given rule. For this chapter's case study example, GGNRA staff investigated the environmental impact of dog walking on plant life and public health.

To share information, agency personnel conduct informal outreach and communication with stakeholders through phone calls and meetings to collect insights regarding how to conceptualize and design regulations that address an issue.[17] These conversations occur one-on-one between agency personnel and an interest representative or an expert in the field under investigation. These conversations are not required by the APA to be included in the public record associated with the rule.

Additionally, agencies can adopt more formal processes to collect information, such as issuing an ANPRM, as noted in the GGNRA case study, or a notice of intent. ANPRMs and notices of intent are formal collection processes for agencies to solicit feedback to ascertain information through the *Federal Register*. Input collected through an ANPRM or other processes does become part of the public record for the rule.

After collecting information, the agency begins to draft rule language. For more salient or controversial rules (e.g., rules with economic significance), agencies pursue a work group approach to develop a proposed rule and its related documents, such as a regulatory impact statement. Although many personnel can be involved in a work group to design a rule and provide input, only a few individuals are tasked with drafting the language of the rule. Agency personnel can also formally include stakeholders in the development of rule language by conducting a reg neg process to develop at least some of the rule language (this process is discussed in more detail later in this chapter).

Review

Once a draft rule is developed, it is submitted for interagency and political appointee review. Many federal agencies have overlapping authority over the environment and the economy. To ensure rules do not negatively impact one another or achieve cross-purposes, agency personnel submit draft rules for interagency review and feedback. Rules also must be reviewed by politically appointed administrators within the agency prior to being proposed to ensure the rule aligns with presidential priorities. For economically significant rules, as discussed later in this chapter, the Office of Management and Budget within the executive office of the president is also required to review draft rules prior to their formal proposal. If a rule clears these review processes, it can then be proposed in the *Federal Register* and enter the notice and comment stage of the process, outlined in Chapter 3.

Rulemaking may start with any one of these preliminary steps in the pre-rule process; relatively few rules make it to the notice and comment stage on a yearly basis due to the amount of time it takes to collect information. Generally, the pre-rule stage is the lengthiest of the rulemaking stages. We delve more deeply into some of the nuances of this stage of the process by exploring fact-finding mechanisms and review processes to understand the rigorous nature of the rule development phase of rulemaking.

Engaging Stakeholders

As the GGNRA case indicates, information sharing is essential for agencies to investigate additional facts to develop a rule. Decades of information sharing occurred in order to eventually propose a dog management rule for the park. Cornelius Kerwin and Scott Furlong recognize interest groups often have access to data or provide technical expertise that agencies need during the pre-proposal stages of a rule. Agencies rarely contain in their possession all the information needed to write a rule.[18]

Incorporating stakeholder feedback during the rule development process has not always been a given for regulatory decision making. Federal agencies and the EPA were criticized for their lack of stakeholder engagement prior to the proposal well into the 1990s.[19] The concern was that agency personnel were developing rules without all the technical resources available from affected stakeholders, resulting in less efficient and more costly rules.[20] This criticism did not go unnoticed by federal agencies, which have experimented with various methods to incorporate more stakeholder input at this formative stage of the process through soliciting comments via informal meetings, ANPRMs, reg neg, and, more recently, reg neg lite (described later in this chapter).

Informal Meetings

Agencies often task one or two individuals with the responsibility to create a rule. These individuals vary in their expertise and backgrounds. For example, many staff with the NPS hold degrees in wildlife biology, law, or restoration science. Although they are experts, their training might not have provided them with all the tools necessary to write a rule on dog management. As a result, park staff might start the process informally, meeting with scientists who have conducted peer-reviewed research on the impacts of dogs on plant life in the San Francisco Bay Area. This information could become valuable for how an agency proceeds. Comparatively, scientific information could lead to an agency deciding to move forward to other pathways for public outreach.

Advanced Notice of Proposed Rulemaking

Public agencies have always had the opportunity to solicit public comment prior to rulemaking via the ANPRM process. Many stakeholders already follow the *Federal Register*, where all proposed and final regulations are published, to identify what rules have been adopted and their impacts on regulated interests. As a result, agencies can conduct outreach to a wide variety of entities via this pathway all at once. Any comment provided during the ANPRM process is then included in the public record about the rulemaking.

Issuing ANPRMs can be time- and resource-intensive, so they are issued sparingly and often only in the most contentious rulemakings. The agency must allocate staff to draft a preamble to the request and identify the areas of interest on which they would like stakeholders to provide feedback, along with processing all the comments received. Even when agencies take the time to publish an ANPRM, they may not receive as much participation as they would like, especially on the specified areas of interest. This is in part because stakeholders may be unwilling to provide technical information they deem confidential or otherwise sensitive in this more formal process. As a result, agency personnel often seek other means to solicit stakeholder feedback at this stage.

Negotiated Rulemaking

Unlike the ANPRM process, in which stakeholders are asked to provide feedback that the agency can use in rule development, in negotiated rulemaking—or reg neg—stakeholders are directly involved in drafting the content of a proposed rule. Reg neg was first introduced in the 1980s to leverage the expertise of stakeholders in rule development to build consensus on rule language and reduce rulemaking time lines and litigation.[21] The expectation was that stakeholder input would result in more effective rules and foster buy-in from affected stakeholders.[22] Initial enthusiasm for reg neg in the 1980s was bolstered by the passage of the Negotiated Rulemaking Act of 1990, which clarified agency authority to conduct reg neg.

The reg neg process is consistent across federal agencies. It is overseen by a nonpartisan convener who sets ground rules. The convener decides whether the reg neg process is feasible for a given rule and then notifies the agency and develops a list of affected parties. From this list of affected parties, a maximum of twenty-five individuals are selected to represent the myriad of affected parties and negotiate the content of the proposed rule. During the negotiation, the affected parties exchange and provide information to explain why a given policy alternative is more feasible or appropriate than others. Ultimately, the goal of the negotiating group is to gain a consensus that a proposed rulemaking is acceptable to all groups. If this consensus is reached, then the group prepares a report containing the agreement regarding the language and content of a proposed regulation and an explanation of how the group came to that decision.[23] Each member is then committed to the agreement and is expected to relinquish his or her litigation rights at the conclusion of the rulemaking process.

Many agencies have used this process to develop rulemakings, as was done by Golden Gate with some positive effects. In some cases, reg neg processes have resulted in stakeholders coming together and agreeing on rule language they could all support.[24] These processes also increased stakeholder perceptions of the legitimacy of the processes, at least for those groups involved in the process.[25]

There are also significant drawbacks to reg neg that have limited its use across the bureaucracy.[26] A key problem for reg neg is that the negotiations are limited to a subset of actors who often must represent a variety of organizations with potentially divergent objectives.[27] This leads to criticism of the process from those entities that are not included in the negotiation.[28] In addition, setting up a reg neg and finding a time when as many as twenty-five organizations can meet to negotiate rule language is a massive undertaking. Finally, reg neg proceedings have only added time to rulemaking proceedings and have not influenced litigation rates.[29] Given these and other factors, few reg neg proceedings have occurred since the 1990s.[30]

Reg Neg Lite

Given the limitations of some rule development pathways, agency personnel have sought other methods to solicit stakeholder feedback during the pre-rule process.[31] Jeffrey Lubbers suggests that some agencies have moved toward "reg neg lite" to gain informal stakeholder input during the pre-proposal stage.[32] Agency personnel can discuss rule language with similar stakeholders instead of bringing together all vested interests, as defined by a formal reg neg. This approach lessens the acrimony or fear that stakeholders' words or information will be attributed back to them, as is done in the ANPRM or reg neg process. In short, reg neg lite allows the agency to generate a larger pool of data from a variety of actors while limiting the time and resources required from both the agency and stakeholders.[33]

Although the technical information provided by stakeholders can result in more effective rules, reg neg lite has its own drawbacks. The interactions are informal, making it difficult to understand which groups provided feedback to the agency and who had an impact over regulatory decisions. In addition, not all stakeholder groups can be contacted informally by the agency. As evidenced by Sara Rinfret and Jeffrey Cook, many may not know the agency is considering a particular rule.[34] In contrast, ANPRM and reg neg processes leave a paper trail that can help stakeholders know where the agency is heading. These issues can result in stakeholders distrusting the outcome of the rule, especially if they were not consulted in its development.

Office of Information and Regulatory Affairs

Agency rule writers sometimes need to closely consider a rule's regulatory impact on the U.S. economy. Economically significant rules (i.e., those that cost more than $100 million) must adhere to Executive Order 12866, which requires the Office of Management and Budget's Office of Information and Regulatory Affairs (OIRA) to review economically significant rules. Additionally, under President Reagan's Executive Order 12291, the OMB director is authorized to "review any draft proposed or final rule or regulatory impact analysis from a covered agency."[35] Any given agency provides OIRA with a cost-benefit analysis for those rules. During the review process, OIRA and the agency communicate about the economic analysis. The gatekeepers that examine the economic significance of the review are desk officers in the OIRA.

When an agency compiles a draft rule, it is submitted through an online portal called ROCIS (RISC/OIRA Combined Information System). Upon submission, the draft rule is forwarded to an OIRA desk officer. The desk officers are trained in economic analysis. Typically, OIRA has forty to forty-five desk officers, and each manages the work, on average, of six or seven specific agencies.[36] For example, one desk officer would oversee public land agencies (e.g., Bureau of Land Management, National Park Service, Fish and Wildlife Service).[37]

The OIRA review process is seen by some as burdensome and as impacting the timeliness of when a rule can move forward to the NPRM stage of the rulemaking process. Under Executive Order 12866, OIRA has 120 days to reject or approve a draft rule. However, Daniel Farber and Anne Joseph O'Connell suggest that OIRA takes much longer for its review and can delay an agency's proposal of an NPRM.[38] Once OIRA does finish its review, an agency is able to decide whether to move forward to the publication of an NPRM for public comment.

Lisa Heinzerling, a former EPA counsel, explains that the lengthy OIRA review process is an outright burden for agencies. From her vantage point, OIRA places an undue burden on agency rule development because the process is fraught with the personal biases of OIRA staff about particular rules. Quite often, OIRA spends a great deal of time reviewing environmental regulations because of concerns regarding monetizing benefits.[39] Lisa Bressman and Michael Vandenbergh, in contrast, provide an agency-centered analysis that examines the perspectives of thirty former EPA presidential appointees and their perceptions of the OIRA regulatory review process. The results of their work suggest the need for the role of accountability in executive-branch decision making due to political bias in the process.[40]

By way of summary, we provide examples of the characteristics driving the rule development phase of the rulemaking process. These range from stakeholder information sharing to OIRA review processes. However, as this explanation demonstrates, outside interests are involved in various points of rule development. For example, interest groups can request meetings with OIRA desk officers to discuss an agency's draft rule. These meetings are posted on OIRA's website for public review. To take a closer look at the role of stakeholders in shaping the rule development phase, we turn to noteworthy scholarship, which extensively investigates whether influence does indeed matter at this stage.

Parsing Influence on Draft Rule Content

Today, federal agencies make concerted efforts to incorporate stakeholder feedback via informal and formal pathways.[41] In the GGNRA case study, consider how much time park personnel spent to reconcile stakeholder differences. As noted, it took fourteen years for park staff to publish an NPRM. This case illuminates how stakeholders can play a fundamental role in shaping the creation of an agency rule. Nonetheless, participating interest groups believe one of the most effective pathways to influence outcomes is involvement during this stage in the process. Yet this begs the question of whether these perceptions match reality: How much influence do these individuals have on agency rule writers during the rule development phase?

As illustrated in the next chapter, the historical focus for scholars is to examine stakeholder influence on the public comment phase during an NPRM.[42] However, scholars note this analysis is incomplete, given that once a rule is drafted, much of the work is already done and agencies are unwilling to make significant changes for fear of litigation (see Chapter 3).[43] As a result, many stakeholders attempt to focus their lobbying efforts during the pre-rule stage to maximize their impact.

Scholars suggest interest groups can have influence at this stage of the process, but which groups and to what extent are unclear.[44] Richard Hoefer and Kristin Ferguson note that access to agency personnel at this stage matters, but they conclude that this access may not be influential. Instead, Rinfret and others contend that it is the types of arguments stakeholders bring to the agency that can result in more (or less) influence on outcomes.[45]

Rinfret, in her examination of the U.S. Fish and Wildlife Service, found that stakeholders often frame their regulatory arguments in three basic categories: expertise, fiscal feasibility, and instructive frames. Arguments associated with the expertise frame relate to scientific or technical opinions about how an agency can address a problem. The fiscal feasibility frame captures arguments that relate to the benefits and costs of possible actions. Finally, the instructive frame encompasses how a stakeholder understands a problem and what the agency should do to solve it. Rinfret's study concludes that interest groups that deploy more arguments from the expertise frame are the most likely to influence outcomes at the pre-proposal stage, and these findings have been corroborated by others.[46] This conclusion makes sense, given that the key goal of agency personnel in seeking stakeholder input throughout the rulemaking process is to collect technical information that can help them develop a more effective rule.

Although these findings are notable, they have generated additional questions: Why do interest groups use other frames when they are ostensibly ineffective? How can we gauge influence when interests provide competing technical arguments? Subsequent work has clarified that interest groups employ other frames to develop a cohesive story that defines the problem, potential solutions, and the preferred outcome.[47] Deploying technical arguments without placing them in the broader context that would justify them is likely to be less effective with agency personnel. This point also provides a pathway for parsing interest group influence across groups. Quite often, interest groups will provide technical arguments to support their preferred approach; however, these arguments are couched within how the interest group defines the issue and the possible outcomes.

If the interest group's instructive frame is far outside what the agency is considering, the group might be unlikely to exert influence regardless of its technical arguments. For example, the EPA may be considering regulation on the air emissions from the idling of passenger vehicles to mitigate local air pollution such as ground-level ozone. One interest group could agree that emissions from vehicles are a problem but argue that emissions can be managed with some improvements to the status quo. Additionally, this group can provide technical information on how a set of new emission control technologies could reduce idling vehicle air emissions. Another interest group may view emissions from idling vehicles as an unacceptable attack on public health. Depending on how agency personnel perceive the problem, one of these two groups may be more likely to influence the outcome of the rule. Thus, it is important to analyze and consider all interest group arguments to determine whether certain groups are more influential on outcomes.

On the question of whether certain groups wield more or less influence over federal agencies at this stage in the process, research is mixed, with some scholars suggesting business interests are dominant and others arguing influence is balanced.[48] Part of the disagreement is associated with the types of rules evaluated: controversial and noncontroversial. In noncontroversial rules, or routine rules where the public is not particularly concerned about the outcome, it appears business interests are the most influential at this stage in the process.[49] This may be in part a function of the content of the rule. Business interests might be impacted by a slight update to a routine rule and so they will participate informally in the process. In comparison, the rule will have a limited effect on the public or the environment, and so other interests may not choose to participate.

For more controversial rules, where federal agencies might make significant changes in policy, evidence suggests that both business and public interest groups choose to engage in the pre-rule stage of the process so they can influence the outcomes through technical arguments. This process is largely driven by the goal of agency personnel to gather as much information as possible while also identifying areas of consensus around rule language during the pre-rule stage. This process can also identify specific areas that might be controversial, allowing the agency to request more information prior to publication of a proposed rule.

Research also suggests that influence is at play during the OIRA review process. For instance, during OIRA's review of an agency rule, stakeholder groups can request meetings with OIRA. Scholars have been interested in determining how influential these groups are in OIRA's determinations of approving or rejecting a proposed rule. Erik Olson and Shelley Lynne Tomkin argue that OIRA is biased in its analysis of rules and favors industrial interests or less regulation overall.[50] This research is supported by Steven Croley, who argues that OIRA also reviews rules at the behest of environmental organizations.[51] Additionally, scholars suggest that OIRA is impacted by partisan leanings or presidential priorities.[52] By way of comparison, Balla, Deets, and Maltzman suggest that OIRA staff under both the Bill Clinton and George W. Bush administrations reviewed more rules after business groups approached OIRA, but the office did not always weaken rules due to these meetings.[53] Former OIRA employee Donald Arbuckle asserts that these arguments are unwarranted, because OIRA is willing and open to meet with anyone during its review process.[54]

The rule development phase is complex and far-reaching. Agencies conduct a great deal of research to best inform the rule. As noted in this chapter, it took several years to propose a rule for dog management in Golden Gate National Recreation Area.

Concluding Thoughts

People often forget about or are unaware of the amount of detail and work that goes into rulemaking processes. Understanding the rule development phase offers a behind-the-scenes perspective, explaining the creation of environmental policy. Rulemaking scholars offer a lens into the involvement of interest groups in the stages of the rulemaking process. As documented here, stakeholders can and do shape the rule development phase.

For example, stakeholders were equally divided in their support of and opposition to pet management in GGNRA. This led to decades of agency rule development. Additionally, since President Reagan, subsequent presidents, regardless of party, have continued the practice of OMB review. In short, the findings presented here illustrate the often overlooked relationships that explain how rules are developed at the federal level. Examining the rule development phase allows us to understand why it can take years for an agency to propose a notice of proposed rulemaking. Agencies spend countless hours sharing information and engaging in outreach to ensure the public has a voice in its processes. Although the APA is silent about documenting this phase of the process, agencies utilize a number of mechanisms to engage the public to participate to inform the next stage of the process.

In the next chapter, Jeffrey Cook uncovers the details of how and why someone would submit a public comment. The public comment phase is ripe for public involvement and invites citizens to understand how to be key players in designing environmental regulations.

Notes

1. National Park Service, "Creation of Golden Gate National Recreation Area," Golden Gate National Recreation Area, California, updated January 30,

2020, https://www.nps.gov/goga/learn/historyculture/creation-of-golden-gate-national-recreation-area.htm.

2. National Park Service, "GGNRA Dog Management Plan," PEPC Planning, Environment and Public Comment, https://parkplanning.nps.gov/projectHome.cfm?projectID=11759.

3. National Park Service, "Golden Gate National Recreation Area Advanced Notice of Proposed Rulemaking (ANPR)," ANPR Overview, https://www.nps.gov/goga/learn/management/upload/goga-anpr_overview.pdf.

4. Jim Burnett, "Efforts to Regulate Off-Leash Dogs at Golden Gate National Recreation Area Spark Debate," *National Parks Traveler*, February 16, 2011, https://www.nationalparkstraveler.org/2011/02/efforts-regulate-leash-dogs-golden-gate-national-recreation-area-spark-debate7613.

5. United States Department of the Interior, National Park Service, "Golden Gate National Recreation Area Advisory Commission, Approved Guidelines for a Pet Policy—San Francisco and Marin County (Muir Beach & South)," February 24, 1979, https://www.nps.gov/goga/planyourvisit/upload/GGNRA-1979-pet-policy.pdf.

6. National Park Service, "Federal Panel Recommendation to the General Superintendent on Proposed Rulemaking for Pet Management at Golden Gate National Recreation Area," revised November 7, 2012, https://www.nps.gov/goga/learn/management/loader.cfm?csModule=security/getfile&PageID=155352.

7. Frederic I. Solop and Kristi K. Hagen, "Golden Gate National Recreation Area: Public Opinion Research Telephone Survey and Public Comment Analysis," National Park Service, August 15, 2002, https://www.nps.gov/goga/learn/management/loader.cfm?csModule=security/getfile&PageID=155357.

8. National Park Service, "Federal Panel Recommendation."

9. Proclamation No. --, 70 Fed. Reg. 123 (June 28, 2005).

10. National Park Service, "NEPA and Negotiated Rulemaking," Golden Gate National Recreation Area, https://parkplanning.nps.gov/document.cfm?parkID=303&projectID=12791&documentID=14267.

11. National Park Service, "Proposed Rule for Dog Management," Golden Gate National Recreation Area, updated September 4, 2016, https://www.nps.gov/goga/getinvolved/prop-rule-dog-mgt.htm.

12. Cary Coglianese and Daniel E. Walters, "Agenda-Setting in the Regulatory State: Theory and Evidence," University of Pennsylvania Law School, Penn Law: Legal Scholarship Repository, 2016, https://scholarship.law.upenn.edu/cgi/viewcontent.cgi?article=2704&context=faculty_scholarship.

13. Office of Information and Regulatory Affairs, "Fall 2019 Unified Agenda of Regulatory and Deregulatory Actions," https://www.reginfo.gov/public/do/eAgendaMain.

14. Coglianese and Walters, "Agenda-Setting in the Regulatory State."

15. Andrew Restuccia, "White House Trumpets Early Success in Wiping Out Regulations," Politico, July 19, 2017, https://www.politico.com/story/2017/07/19/white-house-wiping-out-regulations-240742.

16. The White House, Modernizing Regulatory Review, January 20, 2021, https://www.whitehouse.gov/briefing-room/presidential-actions/2021/01/20/modernizing-regulatory-review/.

17. Cornelius Kerwin and Scott Furlong, *Rulemaking: How Government Agencies Write Law and Make Policy* (Washington, DC: Island Press, 2003).

18. Ibid.

19. Jeffrey J. Cook, "Explaining Innovation: The Environmental Protection Agency, Rule Making, and Stakeholder Engagement," *Environmental Practice* 16, no. 3 (2014): 171–181.

20. Philip J. Harter, "Negotiating Regulations: A Cure for Malaise," *Georgetown Law Journal* 71 (1982): 1–116; Thomas O. McGarity, "Some Thoughts on 'Deossifying' the Rulemaking Process," *Duke Law Journal* 41 (1992): 1385–1462.

21. Harter, "Negotiating Regulations"; Lawrence Susskind and Gerard McMahon, "The Theory and Practice of Negotiated Rulemaking," *Yale Journal on Regulation* 3 (1985): 133–165.

22. Joseph Cooper and William F. West, "Presidential Power and Republican Government: The Theory and Practice of OMB Review of Agency Rules," *Journal of Politics* 50, no. 4 (1998): 864–895.

23. David M. Pritzker and Deborah S. Dalton, "Negotiated Rulemaking Sourcebook," Administrative Conference of the United States, September 1995, https://www.acus.gov/sites/default/files/documents/Reg%20Neg%20Sourcebook%20Chap%201_0.pdf.

24. Susskind and McMahon, "Theory and Practice of Negotiated Rulemaking."

25. Laura I. Langbein and Cornelius M. Kerwin, "Regulatory Negotiation versus Conventional Rule Making: Claims, Counterclaims, and Empirical Evidence," *Journal of Public Administration Research and Theory* 10 (2000): 599–632.

26. Sara R. Rinfret and Jeffrey J. Cook, "Environmental Policy Can Happen: Shuttle Diplomacy and the Reality of Reg Neg Lite," *Environmental Policy and Governance* 24, no. 2 (2014): 122–133.

27. Daniel J. Fiorino, "Regulatory Negotiation as a Policy Process," *Public Administration Review* 48, no. 4 (1988): 764–772.

28. Cary Coglianese, "Assessing Consensus: The Promise and Performance of Negotiated Rulemaking," *Duke Law Journal* 46 (1997): 1255–1348.

29. Ibid.

30. Jeffrey Lubbers, "Achieving Policymaking Consensus: The (Unfortunate) Waning of Negotiated Rulemaking," *South Texas Law Review* 49 (2008): 987–1017.

31. Kerwin and Furlong, *Rulemaking*.

32. Lubbers, "Achieving Policymaking Consensus."

33. Rinfret and Cook, "Environmental Policy Can Happen."

34. Sara R. Rinfret and Jeffrey J. Cook, "Environmental Policy Can Happen: Shuttle Diplomacy and the Reality of Reg Neg Lite," *Environmental Policy and Governance* 24, no. 2 (2014): 122–133.

35. Curtis W. Copeland, "Length of Rule Reviews by the Office of Information and Regulatory Affairs," December 2, 2013, https://www.acus.gov/sites/default

/files/documents/OIRA%20Review%20Final%20Report%20with%20Cover%20Page.pdf.

36. Ibid.

37. Sara Rinfret, "Public Land Agencies, OIRA, and Rulemaking," *Public Administration Quarterly* 41, no. 1 (2017), https://www.questia.com/library/journal/1P3-4316907031/public-land-agencies-oira-and-rulemaking-24.

38. Daniel A. Farber and Anne Joseph O'Connell, "The Lost World of Administrative Law," *Texas Law Review* 92 (2014), https://papers.ssrn.com/sol3/papers.cfm?abstract_id=2395276.

39. Frank Ackerman and Lisa Heinzerling, *Priceless: On Knowing the Price of Everything and the Value of Nothing* (New York: New Press, 2004).

40. Lisa S. Bressman and Michael P. Vandenbergh, "Inside the Administrative State: A Critical Look at the Practice of Presidential Control," *Michigan Law Review* 105, no. 1 (2006), https://repository.law.umich.edu/mlr/vol105/iss1/1/.

41. Jeffrey S. Lubbers, *A Guide to Federal Agency Rulemaking*, 6th ed. (Chicago: American Bar Association, 2002); Kerwin and Furlong, *Rulemaking*.

42. Kerwin and Furlong, *Rulemaking*; Marissa Martino Golden, "Interest Groups in the Rule-Making Process: Who Participates? Whose Voices Get Heard?" *Journal of Public Administration Research and Theory* 8 (1998): 245–270; Jeffrey J. Cook, "Crossing the Influence Gap: Clarifying the Benefits of Earlier Interest Group Involvement in Shaping Regulatory Policy," *Public Administration Quarterly* 42, no. 4 (2018).

43. Jeffrey J. Cook and Sara R. Rinfret, "The EPA Regulates GHG Emissions: Is Anyone Paying Attention?" *Review of Policy Research* 30, no. 3 (2013): 263–280.

44. Richard Hoefer and Kristin Ferguson, "Controlling the Levers of Power: How Advocacy Organizations Affect the Regulation Writing Process," *Journal of Sociology and Social Welfare* 34, no. 1 (2007), https://scholarworks.wmich.edu/jssw/vol34/iss1/6/; William F. West, "Inside the Black Box: The Development of Proposed Rules and the Limits of Procedural Controls," *Administration and Society* 41, no. 5 (2009), https://journals.sagepub.com/doi/10.1177/0095399709339013; Keith Naughton, Celeste Schmid, Susan Webb Yackee, and Xueyong Zhan, "Understanding Commenter Influence during Agency Rule Development," *Journal of Policy Analysis and Management* 28, no. 2 (2009): 258–277, https://onlinelibrary.wiley.com/doi/abs/10.1002/pam.20426; Sara R. Rinfret, "Frames of Influence: U.S. Environmental Rulemaking Case Studies," *Review of Policy Research* 28, no. 3 (2011): 231–246; Susan Webb Yackee, "The Politics of Ex Parte Lobbying: Pre-Proposal Agenda Building and Blocking during Agency Rulemaking," *Journal of Public Administration Research and Theory* 22, no. 2 (2012): 373–393, https://www.researchgate.net/publication/259926296_The_Politics_of_Ex_Parte_Lobbying_Pre-Proposal_Agenda_Building_and_Blocking_during_Agency_Rulemaking.

45. Jeffrey J. Cook, "Are We There Yet? A Roadmap to Understanding National Park Service Rulemaking," *Society and Natural Resources* 27, no. 12 (2014): 1257–1270; Rinfret, "Frames of Influence"; Jeffrey J. Cook, "Framing the Debate:

How Interest Groups Influence Draft Rules at the United States Environmental Protection Agency," *Environmental Policy and Governance* 28 (2018): 183–191.

46. Cook, "Are We There Yet"; Cook, "Framing the Debate."

47. Cook, "Framing the Debate."

48. Wendy Wagner, Katherine Barnes, and Lisa Peters, "Rulemaking in the Shade: An Empirical Study of EPA's Air Toxic Regulations," *Administrative Law Review* 63, no. 1 (2011): 99–158; Jeffrey J. Cook, "Setting the Record Straight: Interest Group Influence on Environmental Protection Agency Climate Policy" (unpublished diss., Colorado State University, 2017), https://dspace.library.colostate.edu/bitstream/handle/10217/181399/Cook_colostate_0053A_14127.pdf?sequence=1; Rinfret, "Frames of Influence."

49. Wagner et al., "Rulemaking in the Shade."

50. Erik D. Olson, "The Quiet Shift of Power: Office of Management and Budget Supervision of Environmental Protection Agency Rulemaking under Executive Order 12,291," *Virginia Journal of Natural Resources Law* 4, no. 1 (1984): 1–80b.; Shelley Lynne Tomkin, *Inside OMB: Politics and Process in the President's Budget Office* (London: M. E. Sharpe, 1998).

51. Steven Croley, "White House Review of Agency Rulemaking: An Empirical Investigation," *University of Chicago Law Review* 70, no. 3 (2003): 821–885.

52. Terry Moe, "The Politicized Presidency," in *The New Direction of American Politics*, ed. John E. Chubb and Paul E. Peterson (Washington, DC: Brookings, 1985).

53. Steven J. Balla, Jennifer M. Deets, and Forrest Maltzman, "Outside Communications and Oira Review of Agency Regulations," *Administrative Law Review* 63 (2011): 149–177, http://www.jstor.org/stable/23065480.

54. Donald R. Arbuckle, "OIRA and Presidential Regulatory Review: A View from Inside the Administrative State," May 3, 2008, http://works.bepress.com/donald_arbuckle/1.

Suggested Readings

Kerwin, Cornelius, and Scott Furlong. *Rulemaking: How Government Agencies Write Law and Make Policy*. 5th ed. Thousand Oaks, CA: CQ Press, 2019.

Rinfret, Sara, and Michelle Pautz. *US Environmental Policy in Action*. 2nd ed. New York: Palgrave Macmillan, 2019.

Rinfret, Sara, Denise Scheberle, and Michelle Pautz. *Public Policy: A Concise Introduction*. Thousand Oaks, CA: CQ Press, 2018.

Bibliography

Ackerman, Frank, and Lisa Heinzerling. *Priceless: On Knowing the Price of Everything and the Value of Nothing*. New York: New Press, 2004.

Arbuckle, Donald R. "OIRA and Presidential Regulatory Review: A View from Inside the Administrative State." May 3, 2008. Available at http://works.bepress.com/donald_arbuckle/1.

Bressman, Lisa S., and Michael P. Vandenbergh. "Inside the Administrative State: A Critical Look at the Practice of Presidential Control." *Michigan Law Review* 105, no. 1 (2006): 47–100. Available at https://repository.law.umich.edu/mlr/vol105/iss1/1/.

Burnett, Jim. "Efforts to Regulate Off-Leash Dogs at Golden Gate National Recreation Area Spark Debate." *National Parks Traveler,* February 16, 2011. Available at https://www.nationalparkstraveler.org/2011/02/efforts-regulate-leash-dogs-golden-gate-national-recreation-area-spark-debate7613.

Coglianese, Cary. "Assessing Consensus: The Promise and Performance of Negotiated Rulemaking." *Duke Law Journal* 46 (1997): 1255–1348.

Coglianese, Cary, and Daniel E. Walters. "Agenda-Setting in the Regulatory State: Theory and Evidence." University of Pennsylvania Law School. Penn Law: Legal Scholarship Repository, 2016. Available at https://scholarship.law.upenn.edu/cgi/viewcontent.cgi?article=2704&context=faculty_scholarship.

Cook, Jeffrey J. "Are We There Yet? A Roadmap to Understanding National Park Service Rulemaking." *Society and Natural Resources* 27, no. 12 (2014): 1257–1270.

———. "Crossing the Influence Gap: Clarifying the Benefits of Earlier Interest Group Involvement in Shaping Regulatory Policy." *Public Administration Quarterly* 42, no. 4 (2018).

———. "Explaining Innovation: The Environmental Protection Agency, Rule Making, and Stakeholder Engagement." *Environmental Practice* 16, no. 3 (2014): 171–181.

———. "Framing the Debate: How Interest Groups Influence Draft Rules at the United States Environmental Protection Agency." *Environmental Policy and Governance* 28 (2018): 183–191.

———. "Setting the Record Straight: Interest Group Influence on Environmental Protection Agency Climate Policy." Unpublished diss., Colorado State University, 2017. Available at https://dspace.library.colostate.edu/bitstream/handle/10217/181399/Cook_colostate_0053A_14127.pdf?sequence=1.

Cook, Jeffrey J., and Sara R. Rinfret. "The EPA Regulates GHG Emissions: Is Anyone Paying Attention?" *Review of Policy Research* 30, no. 3 (2013): 263–280.

Cooper, Joseph, and William F. West. "Presidential Power and Republican Government: The Theory and Practice of OMB Review of Agency Rules." *Journal of Politics* 50, no. 4 (1998): 864–895.

Copeland, Curtis W. "Length of Rule Reviews by the Office of Information and Regulatory Affairs." December 2, 2013. Available at https://www.acus.gov/sites/default/files/documents/OIRA%20Review%20Final%20Report%20with%20Cover%20Page.pdf.

Croley, Steven. "White House Review of Agency Rulemaking: An Empirical Investigation." *University of Chicago Law Review* 70 no. 3 (2003): 821–885.

Farber, Daniel A., and Anne Joseph O'Connell. "The Lost World of Administrative Law." *Texas Law Review* 92 (2014). Available at https://papers.ssrn.com/sol3/papers.cfm?abstract_id=2395276.

Fiorino, Daniel J. "Regulatory Negotiation as a Policy Process." *Public Administration Review* 48, no. 4 (1988): 764–772.

Harter, Philip J. "Negotiating Regulations: A Cure for Malaise." *Georgetown Law Journal* 71 (1982): 1–116.

Hoefer, Richard, and Kristin Ferguson. "Controlling the Levers of Power: How Advocacy Organizations Affect the Regulation Writing Process." *Journal of Sociology and Social Welfare* 34, no. 1 (2007). Available at https://scholar works.wmich.edu/jssw/vol34/iss1/6/.

Kerwin, Cornelius, and Scott Furlong. *Rulemaking: How Government Agencies Write Law and Make Policy*. Washington, DC: Island Press, 2003.

Langbein, Laura I., and Cornelius M. Kerwin. "Regulatory Negotiation versus Conventional Rule Making: Claims, Counterclaims, and Empirical Evidence." *Journal of Public Administration Research and Theory* 10 (2000): 599–632.

Lubbers, Jeffrey S. "Achieving Policymaking Consensus: The (Unfortunate) Waning of Negotiated Rulemaking." *South Texas Law Review* 49 (2008): 987–1017.

———. *A Guide to Federal Agency Rulemaking*. 6th ed. Chicago: American Bar Association, 2002.

Martino Golden, Marissa. "Interest Groups in the Rule-Making Process: Who Participates? Whose Voices Get Heard?" *Journal of Public Administration Research and Theory* 8 (1998): 245–270.

McGarity, Thomas O. "Some Thoughts on 'Deossifying' the Rulemaking Process." *Duke Law Journal* 41 (1992): 1385–1462.

Moe, Terry. "The Politicized Presidency." In *The New Direction in American Politics*, edited by John E. Chubb and Paul E. Peterson. Washington, DC: Brookings, 1985.

National Park Service. "Creation of Golden Gate National Recreation Area." Golden Gate National Recreation Area, California. Updated January 30, 2020. Available at https://www.nps.gov/goga/learn/historyculture/creation -of-golden-gate-national-recreation-area.htm.

———. "Federal Panel Recommendation to the General Superintendent on Proposed Rulemaking for Pet Management at Golden Gate National Recreation Area." Revised November 7, 2012. Available at https://www.nps.gov/goga/learn /management/loader.cfm?csModule=security/getfile&PageID=155352.

———. "GGNRA Dog Management Plan." PEPC Planning, Environment and Public Comment. Available at https://parkplanning.nps.gov/projectHome .cfm?projectID=11759.

———. "Golden Gate National Recreation Area Advanced Notice of Proposed Rulemaking (ANPR)." ANPR Overview. Available at https://www.nps.gov /goga/learn/management/upload/goga-anpr_overview.pdf.

———. "NEPA and Negotiated Rulemaking." Golden Gate National Recreation Area. Available at https://parkplanning.nps.gov/document.cfm?parkID= 303&projectID=12791&documentID=14267.

————. "Proposed Rule for Dog Management." Golden Gate National Recreation Area. Updated September 4, 2016. Available at https://www.nps.gov/goga /getinvolved/prop-rule-dog-mgt.htm.

Naughton, Keith, Celeste Schmid, Susan Webb Yackee, and Xueyong Zhan. "Understanding Commenter Influence during Agency Rule Development." *Journal of Policy Analysis and Management* 28, no. 2 (2009): 258–277. Available at https://onlinelibrary.wiley.com/doi/abs/10.1002/pam.20426.

Office of Information and Regulatory Affairs. "Fall 2019 Unified Agenda of Regulatory and Deregulatory Actions." Available at https://www.reginfo.gov/public /do/eAgendaMain.

Olson, Erik D. "The Quiet Shift of Power: Office of Management and Budget Supervision of Environmental Protection Agency Rulemaking under Executive Order 12291." *Virginia Journal of Natural Resources Law* 4, no. 1 (1984): 1–80b.

Pritzker, David M., and Deborah S. *Dalton*. "Negotiated Rulemaking Sourcebook." Administrative Conference of the United States. September 1995. Available at https://www.acus.gov/sites/default/files/documents/Reg%20Neg %20Sourcebook%20Chap%201_0.pdf.

Proclamation No. --, 70 Fed. Reg. 123 (June 28, 2005).

Rinfret, Sara R. "Frames of Influence: U.S. Environmental Rulemaking Case Studies." *Review of Policy Research*, 28, no. 3 (2011): 231–246.

————. "Public Land Agencies, OIRA, and Rulemaking." *Public Administration Quarterly* 41, no. 1 (2017). Available at https://www.questia.com/library /journal/1P3-4316907031/public-land-agencies-oira-and-rulemaking-24.

Rinfret, Sara R., and Jeffrey J. Cook. "Environmental Policy Can Happen: Shuttle Diplomacy and the Reality of Reg Neg Lite." *Environmental Policy and Governance* 24, no. 2 (2014): 122–133.

Solop, Frederic I., and Kristi K. Hagen. "Golden Gate National Recreation Area: Public Opinion Research Telephone Survey and Public Comment Analysis." National Park Service. August 15, 2002. Available at https://www.nps.gov /goga/learn/management/loader.cfm?csModule=security/getfile&PageID =155357.

Susskind, Lawrence, and Gerard McMahon. "The Theory and Practice of Negotiated Rulemaking." *Yale Journal on Regulation* 3 (1985): 133–165.

Tomkin, Shelley Lynne. *Inside OMB: Politics and Process in the President's Budget Office*. London: M. E. Sharpe, 1998.

United States Department of the Interior, National Park Service. "Golden Gate National Recreation Area Advisory Commission: Approved Guidelines for a Pet Policy—San Francisco and Marin County (Muir Beach & South)." February 24, 1979. Available at https://www.nps.gov/goga/planyourvisit/upload /GGNRA-1979-pet-policy.pdf.

Wagner, Wendy, Katherine Barnes, and Lisa Peters. "Rulemaking in the Shade: An Empirical Study of EPA's Air Toxic Regulations." *Administrative Law Review* 63, no. 1 (2011): 99–158.

Webb Yackee, Susan. "The Politics of Ex Parte Lobbying: Pre-Proposal Agenda Building and Blocking during Agency Rulemaking." *Journal of Public Administration Research and* Theory 22, no. 2 (2012): 373–393. Available at https://www.researchgate.net/publication/259926296_The_Politics_of_Ex _Parte_Lobbying_Pre-Proposal_Agenda_Building_and_Blocking_during _Agency_Rulemaking.

West, William F. "Inside the Black Box: The Development of Proposed Rules and the Limits of Procedural Controls." *Administration and Society* 41, no. 5 (2009): 576–599. Available at https://journals.sagepub.com/doi/10.1177 /0095399709339013.

3

Notice and Comment

What Does the Public Think?

Jeffrey J. Cook

In 2017, comedian John Oliver, the host of *Last Week Tonight*, included a segment about the Federal Communications Commission's (FCC) proposed Net Neutrality Rule.[1] In this controversial regulation, the FCC planned to repeal its existing net neutrality rules that had required Internet service providers to offer equal access to all web content. If the rules were repealed, Oliver feared that Internet service providers would charge more for access to certain content or for higher-speed connections to certain pages.

Oliver called on viewers to visit a website called GoFCCyourself.com that his team had developed. The site redirected viewers to the FCC public comment portal, where they could urge the commission to retain strong net neutrality rules.[2] The first day after the episode aired, the FCC received 150,000 comments, despite the website crashing due to the flood of public comments.[3] Over the course of the comment period, the FCC reported receiving more than twenty-two million comments—a historic total.[4]

This example demonstrates the growing ease of submitting public comments to federal agencies. Almost gone are the days of submitting written comments through the mail. Today, people can submit comments from the comfort of their homes with a few mouse clicks. However, submitting a public comment does not always translate into influence. In the net neutrality example, the FCC ultimately went through with the repeal, despite most commenters opposing the action.[5]

Unlike Congress, federal agencies do not necessarily follow public opinion and are unlikely to make changes to regulatory policy simply because commenters think the rule is a good or bad idea. Rather, agencies tend to make changes based on the technical substance of comments—or comments that demonstrate *why* an agency should change its decision.[6]

Even so, public agencies often do make changes to rules, and so it is imperative to understand how the process works, why agencies make changes, and in response to which commenters. This chapter begins with a look at the origins of the notice and comment period and its evolution to the present day. Next, I highlight the types of interests that are most likely to participate in the notice and comment process, including a case study that helps identify why some interests are more impactful than others—namely, those with relevant technical expertise, a strong coalition, and sustained participation. Finally, the chapter ends with a discussion of why interests participate at this stage of the rulemaking process, even if significant rule changes are unlikely.

How the Process Works

The passage of the Administrative Procedure Act (APA) in 1946 provided Oliver the platform to pull off his ultimately unsuccessful FCC comment campaign. As discussed in Chapter 1, the APA requires agencies to provide notice of a proposed rulemaking and offer the public an opportunity to comment on that proposal. Today, federal agencies are more transparent about the regulatory actions they are taking prior to proposal (see Chapter 2), but the notice and comment stage still provides a critical opportunity for the general public to participate in the rulemaking process.

The process begins when the agency publishes a notice of proposed rulemaking (NPRM) in the *Federal Register.* The APA is silent on the length of the required comment period, but it typically lasts between thirty and ninety days. Regulations that are more complex or controversial might be given a longer commenting period, while relatively routine rules may receive a shorter time line. Regardless of the length of the commenting period, commenters can request extensions, and the agency has the authority to grant or deny these extensions.

There are three key pathways to provide feedback during the notice and comment stage: submit a comment through the mail, email or electronically upload a comment, and provide oral feedback at public hearings. Historically, comments were submitted in person at the agency or through the mail. Today, most comments are delivered electronically via email or uploaded directly to an agency via regulations.gov. Though not required by the APA, agencies often hold hearings for the public to provide oral feedback.[7] These hearings can be held at the location of the agency's discretion. Most agencies are headquartered in the District of Columbia, but it is common for agencies to hold public hearings that are geographically dispersed or in locations that are convenient for a critical mass of stakeholders.[8] All this feedback is then entered into the public record for the rule. These three pathways are discussed in more detail later in this chapter.

Once the comment period closes, the most difficult component of the process begins—comment review. Most regulatory actions, including those in the environmental context, are considered routine and highly technical. These rules typically receive fifty or fewer comments.[9] In comparison, a handful of rules across the bureaucracy are controversial, such as the Obama administration's climate change regulations and the aforementioned FCC rule. These rules can receive hundreds of thousands—and in some cases, millions—of comments.[10] First-of-a-kind rulemakings or rules that result in significant changes in policy often become controversial. These rules spur stakeholder interest and related comments despite the technical nature of the rule.[11] Regardless of the number of comments received, the APA requires the agency to consider all comments in the development of a final rule.

Career civil servants begin the comment review process by first organizing the comments and identifying unique comments as compared with duplicative or substantially similar comments. Substantially similar comments are often associated with interest group–driven mass comment campaigns, as exemplified by Oliver's efforts in the FCC Net Neutrality Rule. These similar comments are batched together for the review process. Agency personnel divide up the unique comments and analyze the content of each unique letter. These personnel then provide a response to the substance of the comment, which is included in a response-to-comment document that may run to hundreds of pages depending on the number of comments received.[12]

After comment review, the agency often makes changes to the rule. The Obama administration spent years crafting its proposed Clean Power Plan—a plan to reduce carbon emissions from existing power plants. The Environmental Protection Agency (EPA) then made many changes to the rule in response to the comments. For example, the agency reduced the expected emissions savings that could be generated by improvements in heat rate at coal-fired power plants in response to the technical data provided via public comments.[13] Career civil servants are the workhorses in recommending rule changes, but any decision to alter rules requires the input and approval from politically appointed administrators in the agency and the Office of Management and Budget (OMB) within the executive office of the president.[14]

After the review of public comments is complete, the agency must determine whether the rule is ready for final publication. At this juncture, the agency has three options: publish, withdraw, or cancel the rule. As noted, the FCC finalized the Net Neutrality Rule in 2017, as did the EPA with the Clean Power Plan in 2015. In some cases, the agency might review the comments and determine that it is not ready to publish the proposed rule in its current form. In this case, the agency can withdraw the rule to make changes to the rule and then reopen it for public comments. This is the approach the EPA took for its New Source Performance Standards for carbon emissions from new power plants; the standards were initially proposed in 2012 and were withdrawn in 2014. The agency revised, proposed, and finalized the rule in 2015.[15] In rare cases, an agency decides to cancel a rule after reviewing the comments. The Obama administration took this step, canceling a proposed ozone rule with an estimated $90 billion in compliance costs in 2011.[16] In this controversial decision, the administration withdrew the rule in line with its broader effort to support the recovering economy.

The agency usually publishes a final rule with at least some changes associated with public feedback. Once the final rule is published, it moves to stage three, the final stage of the process (see Chapters 1 and 4).

Key Functions

Congress adopted the APA to instill democratic norms in the rule-making process, but the law also serves other important functions. The federal government is not the only entity that employs technical experts in each field. Moreover, the best ideas to solve a problem do not always come from behind a government desk. Providing a comment period provides the agency an opportunity to solicit feedback from the public and other technical experts.

The comment period also provides the agency a moment to press the pause button on the regulatory process and allow for another critical objective—proofreading. The public and, more commonly, the regulated community can read the rule line by line and identify errors in regulatory requirements, conflicting requirements, or confusing language. The regulated community wants to ensure they know what a rule covers and, more importantly, how they can comply. Agencies often make changes that tighten rule language in response to public comments, ensuring that the agency publishes a final rule that achieves the goals it is supposed to.

Finally, the notice and comment period provides stakeholders with a formal announcement that the agency is planning to change regulatory requirements. Even though agencies today typically conduct outreach prior to the comment period (see Chapter 2), they cannot reach out to every entity impacted by a rule. Moreover, regulated interests and the public do not always have the resources to follow the rulemaking agenda of every agency, or even departments within the same agency. Thus, an NPRM in the *Federal Register* serves a critical function as the basis for all stakeholders to learn that new regulatory requirements are coming. This can provide stakeholders with the lead time to review the requirements and understand what is needed to demonstrate compliance, allowing them to more effectively comply with the final rule. Federal agencies have taken a variety of pathways to ensure public participation at this stage in the process.

Evolution of the Process

For decades, members of the public who wished to submit comments had to mail or hand-deliver a letter to the agency within the designated comment period. This approach had a constraining impact on public comments, because interested stakeholders had to review the *Federal Register* to identify that a proposed rule was announced, take the time to draft a public comment, and pay sufficient postage to mail the letter and have it arrive before the comment period closed. Although the business community, which is subject to regulatory compliance, has a strong incentive to track the *Federal Register* and submit public comments, the general public is less likely to participate given these inconveniences. Regulation carries direct compliance costs for businesses, while benefits and costs to the public are widely distributed and can be hard to quantify.[17]

The development of the Internet in the 1990s revolutionized the public comment period, as agencies began to allow the public to submit comments electronically.[18] Congress passed the E-Government Act of 2002 with three key goals: (1) increase public access to rulemaking materials, (2) increase public participation, and (3) improve rulemaking efficiency and effectiveness. The law encouraged all agencies to accept electronic comments and established the eRulemaking Program Management Office. Housed within the EPA, the office's objective is to develop and maintain a centralized rulemaking docket system to track all proposed rules and associated comments. The regulations.gov website was launched in 2003, and forty federal agencies and departments serve on the executive committee and in technical work groups.[19]

For participating agencies, regulations.gov has served its goal of providing the public increased access to rulemaking materials, including proposed rules, regulatory impact statements, and the public comments submitted in response to a proposal. In addition, public participation in rulemakings has substantially increased since the website was launched.[20]

E-rulemaking's performance in improving rulemaking efficiency and effectiveness is more mixed. Initially, scholars argued that e-rulemaking would encourage more public interest groups to participate in rulemaking processes, providing a different perspective on regulations than the business community.[21] This diversity of perspectives was expected to make rulemaking processes more deliberative and thereby improve the quality of rulemaking.[22] It was also expected to enhance the legitimacy of rules, given that the agency would make changes to the rulemaking in response to a broader array of actors who are brought into the process and accept the outcome.[23] This buy-in would result in fewer stakeholders filing litigation after a rule is adopted, reducing regulatory costs and shortening time lines.

Although agencies are receiving more public comments, the diversity of commenters as well as the impact of these comments on outcomes is questionable.[24] E-rulemaking has encouraged interest groups of all types to conduct mass comment campaigns in which an interest group solicits its members to submit the same (or essentially similar) "postcard" comment in support of or opposition to a rule. These mass comment campaigns can result in millions of comments submitted for a particular rule, including the aforementioned FCC Net Neutrality Rule.

Many agencies beyond the FCC receive many more public comments, but e-rulemaking and related mass comment campaigns are not increasing input from underrepresented groups.[25] In contrast, agencies are receiving a greater number of duplicate perspectives—from business, environmental, and other public interest groups that already participated prior to e-rulemaking. At the same time, underrepresented groups such as the poor and minority interests continue to lack representation.

Mass comment campaigns also do not provide substantive or technical feedback on the proposed rule; rather they offer only a thumbs-up or thumbs-down on the agency's regulatory decision. As a result, these comments do not provide varying perspectives on how to address the intent of the regulation. Instead, they offer a policy-driven opinion on the action itself that is discounted by agencies given the lack of technical substance.[26] As a result, mass comment campaigns fail to build legitimacy or buy-in across stakeholders, and they do not decrease litigation activities.[27]

The implementation of e-rulemaking has resulted in unintended consequences. It places a strain on agency personnel who must review all the submitted comments that may or may not provide technical information on a rule. In addition, allowing for the submission of electronic comments has increased the potential for comment fraud.[28] Nefarious individuals (domestic or foreign) can use a software bot to submit comments from fake email accounts or to impersonate real people. Agencies might perceive these to be unique comments when, in fact, they were all submitted by the same individual. This type of foreign interference in rulemaking is especially problematic, as exemplified by the estimated half million comments submitted from Russian accounts in the FCC Net Neutrality Rule. The initial goal of e-rulemaking was to enhancing stakeholders' perceptions of rulemaking validity, but comment fraud has sowed doubt about the legitimacy of regulatory processes and has negatively impacted rulemaking.[29]

Coping Strategies

Agency personnel have developed strategies to cope with these unintended consequences. To address comment load, agency personnel have long used various software programs to organize comments. This software can help agencies streamline their review of submissions associated with mass comment campaigns, allowing agency personnel to focus on the unique comments submitted.[30]

Agencies are just beginning to grapple with how to respond to the submission of fake comments, especially from foreign actors. Though opportunities exist to reduce commenter fraud, including identity verification, these steps could both discourage real citizens from participating and add costs to the process.[31] In 2018, the U.S. Government Accountability Office launched an investigation into fake comments submitted in the FCC's Net Neutrality Rule, which may provide additional insights regarding the FCC's handling of the comments and opportunities to address fraud in subsequent rulemakings.[32] In sum, the E-Government Act of 2002 has not met all its expectations, but Congress is unlikely to repeal it. Going forward, agencies will need to continue to find ways to encourage participation from U.S. citizens.

The Participants

A key factor that hinders public participation is that submitting a public comment either online or through the mail is a nontrivial task. Individuals are more likely to submit a public comment if they feel they are personally affected by the proposal. In short, if individuals believe a rule will be detrimental to their way of life, they are more likely to submit a comment. This perspective helps explain why business interests remain the most likely to substantively comment on a rulemaking, because new regulatory requirements can add substantial costs to business activities.[33] For example, the EPA adopted a new Mercury and Air Toxics Standards (MATS) in 2011 with the goal of reducing mercury emissions from fossil fuel–fired power plants.[34] The agency estimated the rule would provide billions of dollars in benefits, including avoiding eleven thousand premature deaths, but these benefits came with an annual compliance cost of $7 billion to $10 billion. At the time, these compliance costs made the rule the most expensive ever adopted by the agency.[35] Although many of these costs are passed on to consumers, the regulations can fundamentally impact the markets in which businesses operate and potentially require costly new business investments to remain competitive.

As the MATS example suggests, businesses are not the only entities impacted by rules; the general public is affected as well. A key goal of environmental regulation is to maintain clean air, water, and land, and these decisions are frequently correlated with public health outcomes. As was done in MATS, public health benefits, such as avoided deaths, are quantified and compared with compliance costs in the justification of proposed rules. Therefore, individuals and public interest groups are frequent participants in regulatory proceedings, because agency decisions can have real effects on people's lives. Even so, public health and environmental benefits are more diffuse, while compliance costs can be concentrated across a few actors. Therefore, business interests are more likely to participate in a rulemaking than are members of the public who may be impacted by the action.

Although business interests are more likely to submit public comments, public interests do participate. The proliferation of public interest groups in the 1970s correlates with increases in public input on policy broadly.[36] Moreover, because e-rulemaking has made it even easier to submit public comments, more members of the public have participated.

The Influencers

Simply participating in a rulemaking does not always result in influencing the outcome. Recall that agency personnel must only consider the public comments when finalizing a proposed rule, not necessarily adopt proposed changes. Rulemaking participants must convince agency personnel that a rule change is needed based on the technical substance of their comment. Despite these constraints, agencies often do make changes in response to the comments received.[37]

Scholars have concluded that public agencies are more likely to make rule changes in response to a business interest commenter.[38] This might suggest public agencies are beholden to the interests they regulate, but reality offers a more nuanced perspective. First, business interests do not represent a united front in regulatory proceedings.[39] Businesses compete against one another to serve customers, and organizations often have different competitive advantages and goals. For example, some car manufacturers such as Tesla develop electric-only vehicles. General Motors manufactures more conventional combustion engine vehicles. These interests can have varying regulatory objectives, despite both being part of the automotive industry.[40]

Although agencies tend to make changes to rulemakings in response to at least some business interests, the changes are often minor clarifying details of the rule.[41] A key reason behind an agency's reluctance to grant significant rule changes is that the courts might throw out the rule if it looks dramatically different from the proposal. Once a rule gets to the notice and comment stage, years of work to conceptualize and draft the rule have been invested in the process. If the agency publishes a rule that is dramatically different from the proposal, the courts might throw out the rule on procedural grounds and require a new regulatory proceeding. Given that regulated businesses are keenly interested in both the overall scope of the rule and specific compliance details, these stakeholders may be more likely to ask the agency to clarify the scope of the rule or to simplify the compliance requirements. This tracks with the reality that business interests are frequently granted many rule changes, but the changes typically do not influence the intent or overall scope of the regulation.

Business interests may receive the most rule changes, but they are not the only group that wins them. Environmental and other interests can also be granted rule changes in response to their technical comments, suggesting the process is not purely dominated by business interests.[42]

Finally, the notice and comment period does not occur within a policy vacuum. The scope of influence of various actors is informed by these and competing interests' activities earlier in the process.[43] Those interests that are involved in the pre-proposal stage of the process have a higher probability of further influencing the rule during the notice and comment stage than those interests that do not participate earlier in the process.[44] These groups are best positioned to influence the agency, because they have a better understanding of areas where the agency might be more willing to compromise on rule language while building trust between the interest group and the agency. This chapter's climate change case study situates these findings in a real-world example.

Case Study: Why Do Some Stakeholders Win Rule Changes While Others Do Not?

Thousands—and in some cases, millions—of people comment on rules, but only a few commenters are granted rule changes. This case study summarizes the notice and comment stage of two climate change rulemakings to contextualize interest group impact and understand why some interests are influential and others are not.

Climate change was a key component of the Obama administration's environmental agenda, and the EPA was the key regulatory agency implementing this controversial agenda.[45] The focus here is on interest group influence on two of the EPA's climate rules: the Prevention of Significant Deterioration and Title V Greenhouse Gas Tailoring Rule (Tailoring Rule) and the 2017 Model Year Light-Duty Vehicle Greenhouse Gas Emissions Rule (2012 CAFE Standards Rule).

The EPA proposed the Tailoring Rule on October 27, 2009, to require large point sources of pollution (i.e., power plants and heavy manufacturing facilities) to adopt best available control technology for greenhouse gas emissions when building new or modifying existing facilities.[46] In short, any facility that emitted more than ten thousand tons per year of carbon dioxide would be required to install best available control technology when making any modifications at an existing or new facility. The agency accepted comment on this proposal until December 28, 2009, and held two public hearings on the proposal. The agency received more than 446,000 comments on the rule. In response to the comments, the EPA made fifteen changes to the rule, including increasing the emissions threshold to twenty-five thousand tons from the previous ten thousand tons.[47]

The EPA proposed the 2012 CAFE Standards Rule on December 1, 2011.[48] This rule required passenger vehicle manufacturers to meet a fleetwide fuel economy average of 54.5 miles per gallon by 2025. The proposed rule offered a range of pathways to achieve that target beyond just fuel economy improvements. The agency accepted comment on the proposal until January 30, 2012, and held three public hearings across the country. The agency received over 300,000 comments on the proposal and made twenty-nine changes to the rule, largely focused on modifications to the compliance credits, including engine stop/start technology.[49]

Assessing Commenter Impact

After filtering out duplicate mass comment letters, the EPA received 326 unique comments in the Tailoring Rule and 140 in the 2012 CAFE Standards Rule.[50] A total of seventy-eight commenters were credited with requesting at least one of the fifteen rule changes in the Tailoring Rule, while forty-one commenters were credited in at least one of the twenty-nine changes in that rule. This demonstrates that between 25 and 30 percent of the unique commenters were credited with at least one rule change. As the literature suggests, business interests were more likely to receive rule changes. These interests were credited with six of the fifteen and twenty-four of the twenty-nine rule changes, respectively. In the remaining rule changes, some combination of interests—state interests or other interests (i.e., public interest organizations)—were credited in the agency's decision to alter the proposed rule.[51] For both rules, most commenters did not receive a rule change, so what factors help explain those that did?

Technical Expertise

The EPA made changes to each rule based on the technical substance of the comments received. In no situation did the EPA highlight a mass comment letter as justification for making a rule change. Commenters who leveraged their technical expertise to persuade the agency to clarify regulatory intent or address the details of the rule as opposed to the overall scope or approach were more successful. For example, commenters in the 2012 CAFE Standards Rule were granted several rule changes requesting clarification of how the agency completed certain regulatory analyses, which vehicles are eligible for certain credits, and how certain credits can be applied.[52]

Coalition Building

Combining technical expertise via coalition building was another means to entice the agency to make a rule change.[53] Here, the concept of the squeaky wheel getting the grease is at play. If more interests suggest a rule change is necessary, it gives the agency pause both to review that requirement and to consider the alternatives proposed by the coalition.[54] This approach can also help interests show a more united front to the agency. This was a key factor in the agency adopting a higher emission threshold in the Tailoring Rule, with a combination of fifty-four business and twelve state interests requesting the same increase.[55]

Participation—Early and Often

Finally, interest groups that were involved earlier in the process were more effective at influencing the agency than those that only submit public comments.[56] This suggests that these interest groups better understand where the agency is confident in its proposed actions along with the areas that may still be up for debate. Commenters who focus their time and resources on the areas where the agency is still formulating its position are more likely to influence the outcome.

This analysis demonstrates that a variety of interest groups are granted rule changes in response to the notice and comment period. Interests that have relevant technical expertise, a strong coalition, and sustained participation throughout the rulemaking process are most likely to receive a rule change. Ultimately, understanding why an agency might have granted a rule change in response to a comment requires an analysis of the specific content of the comment. These indicators can serve as a foundation for conducting that assessment.

Reframing the Role of the Notice and Comment Stage

The passage of the APA was transformative because it represented the first time agencies were required to solicit public input during rule development. Many entities now participate in the notice and comment process, but few receive a rule change. Those revisions that are granted often do not dramatically alter the outcome or scope of a rule. However, I offer three key justifications for why this stage is essential to the process: (1) details matter, (2) political cover, and (3) legal standing.

First, the scope of impact associated with a rule change is in the eye of the beholder. Simply stated, just because scholars argue a rule has not undergone "significant" change from the proposal to the final version does not mean that the changes were not important to at least some people. For example, let's assume you own an oil and gas company with seven wells. The EPA then proposes a rule to cut air pollution emissions at all midsize and large oil and gas companies' wells, requiring new emission control technologies. The EPA sets the threshold for a midsize company at five wells. Your company submits a comment and argues that your firm is inappropriately considered a midsize company. Based on your data, the EPA agrees and confirms the intent of the regulation was misapplied and subsequently finalizes the rule to apply to companies with ten or more wells. This change provides real benefits to your company and others that are similarly situated while not significantly altering the rule's overall impact on emission reduction.

This logic does not track to mass comment campaigns. Rarely do federal agencies reference mass comment campaigns as a justification for rule changes, while mass comment campaigns come with real costs. Interest groups have a limited membership and have competing priorities, so groups must be strategic in how they leverage their membership to maximize their impact. If mass comment campaigns rarely produce rule changes, why do interest groups spend precious staff time organizing them and mobilizing members when other, ostensibly more effective, lobbying strategies could be deployed?

This leads to the second important element of the notice and comment period: political cover. Environmental and supportive public interest groups were some of the first organizations to pursue mass comment campaigns, and the goal was to show support for or opposition to an agency's proposed action. Subsequently, business and business-friendly public interest groups have supported their own mass comment campaigns to serve as a counterweight to environmental interests.[57] Although agencies rarely make changes in response to mass comment campaigns, they do reference the aggregate view of the public comments.[58] If more people are in favor of the rule, the agency frequently mentions that as a justification for completing the action. In those cases when regulatory decisions are not viewed favorably, public agencies can feel pressure to consider alternative options. In these cases, it is also more likely elected officials will exert their own pressure on an agency to reconsider its position. As a result, these comment campaigns may not result in a direct rule change, but they can be an important political tool for or against the agency action beyond the notice and comment stage (see Chapter 4).

Finally, many interest groups use their participation in the notice and comment stage for another means—legal standing. These entities will foreshadow the legal challenges they might pursue against the agency if the agency pursues or withdraws a proposed rule. If the agency reviews the interest group's comment and decides to disregard the group's reasoning, the agency must include that in the public record. The interest group can then file litigation at the conclusion of the process to take its case to the courts. As a result, some interest group comments focus exclusively on legal arguments as opposed to providing technical feedback on how to implement the agency's proposed action. An analysis of the interest group's activities during the notice and comment stage might reveal that the group did not influence the rule, while in reality the same group's public comment could have fundamentally altered the final outcome of the rule, assuming the court invalidates the rule in response to the litigating interest group.

Concluding Thoughts

The notice and comment stage of the rulemaking process is consequential. And in the era of e-governance, it is necessary to explore and consider such topics as expanding access and addressing cybersecurity threats.

A key factor for the lack of public participation is that agencies look for technical expertise and data-driven arguments to justify potential rule changes. Although relatively few people have advanced degrees in the requisite fields to provide this technical feedback, many members of the public have place-based or situational knowledge that could provide agencies with insights on how to more effectively achieve regulatory goals.[59] Agency personnel need to continue to find ways to educate the public about the rulemakings they undertake while allowing the public the opportunity to provide this useful feedback. For example, public agencies could pursue more outreach via social media to solicit interest in agency rulemakings and identify pathways to incorporate the public's feedback. Agencies could also rely on video-chat applications to solicit oral feedback from stakeholders who are unable to attend public hearings. Through these and other pathways, stakeholders could more easily offer their situational knowledge, potentially from impacted sites in real time.

Any effort to expand stakeholder access to the rulemaking process, especially via online or mobile means, will increase threats to cybersecurity, including enabling foreign and nefarious commenting. Developing strategies to verify U.S. citizens who provide their expertise in rulemakings will be essential to maintain the legitimacy of the process. Verification will require public agencies to adopt technologies that are already available, including Completely Automated Public Turing tests to tell Computers and Humans Apart (CAPTCHA), applications that require completion of a series of tasks to verify that an individual is not a bot.[60] At the same time, agency personnel need to consider the additional time and resource burdens verification can bestow on the public. If commenting becomes more arduous, the public is even less likely to participate.

Future rule writers will have to navigate these and other unforeseen challenges to maintain or rebuild confidence that regulatory policy decisions are based on sound, technical grounds. This is a tall task, but academic institutions have developed programs to support the next generation of rule writers, such as the University of Pennsylvania Program on Regulation and the Regulatory Policy Program at Harvard University's Kennedy School of Government.[61]

Even so, when rules are more controversial (i.e., they have economic significance), political institutions (e.g., the president, Congress, and the courts) are more likely to get involved to alter the course of final rules and the activities of an agency more broadly.[62] Whether these institutions become involved varies by topic and agency. These and other implications of political oversight are addressed in Chapter 4.

Notes

1. Melissa Locker, "John Oliver Wants You to Flood the FCC Website to Save Net Neutrality, Again," *Time*, May 8, 2017, https://time.com/4770205/john-oliver-fcc-net-neutrality/.

2. Ibid.

3. Ted Johnson, "FCC Inspector General: John Oliver Segment Triggered System Slowdown, Not Bots," *Variety*, August 7, 2018, https://variety.com/2018/politics/news/john-oliver-fcc-triggered-slowdown-1202898328/.

4. Glenn Fleishman, "FCC Chair Ajit Pai Admits Millions of Russian and Fake Comments Distorted Net Neutrality Repeal," *Fortune*, December 5, 2018, http://fortune.com/2018/12/05/fcc-fraud-comments-chair-admits/.

5. Keith Collins, "Net Neutrality Has Officially Been Repealed. Here's How That Could Affect You," *New York Times*, June 11, 2018, https://www.nytimes.com/2018/06/11/technology/net-neutrality-repeal.html.

6. Jeffrey J. Cook, "Crossing the Influence Gap: Clarifying the Benefits of Earlier Interest Group Involvement in Shaping Regulatory Policy," *Public Administration Quarterly* 42, no. 4 (2018).

7. Cornelius M. Kerwin and Scott Furlong, *Rulemaking: How Government Agencies Write Law and Make Policy*, 3rd ed. (Washington, DC: CQ Press, 2018).

8. For example, the Environmental Protection Agency and the National Highway Traffic Safety Administration held three public hearings on the Safer Affordable Fuel Efficient Vehicles Rule in 2018, with one being located in Dearborn, Michigan, where Ford and other large domestic car manufacturers have significant operations. See United States Environmental Protection Agency, "The Safer Affordable Fuel Efficient (SAFE) Vehicles Proposed Rule for Model Years 2021–2026," September 27, 2018, https://www.epa.gov/regulations-emissions-vehicles-and-engines/safer-affordable-fuel-efficient-safe-vehicles-proposed.

9. Wendy Wagner, Katherine Barnes, and Lisa Peters, "Rulemaking in the Shade: An Empirical Study of EPA's Air Toxic Regulations," *Administrative Law Review* 63, no. 1 (2011): 99–158.

10. Jeffrey J. Cook, "Setting the Record Straight: Interest Group Influence on Environmental Protection Agency Climate Policy" (unpublished dissertation, Colorado State University, 2017), https://dspace.library.colostate.edu/bitstream/handle/10217/181399/Cook_colostate_0053A_14127.pdf?sequence=1; Johnson, "FCC Inspector General."

11. Mark Allen Eisner, Jeffrey Worsham, and Evan J. Ringquist, *Contemporary Regulatory Policy* (London: Lynne Rienner, 2006).

12. Cook, "Setting the Record Straight."

13. United States Environmental Protection Agency, "Fact Sheet: Clean Power Plan Key Changes and Improvements: From Proposal to Final," last modified May 9, 2017, https://archive.epa.gov/epa/cleanpowerplan/fact-sheet-clean-power-plan-key-changes-and-improvements.html.

14. Simon F. Haeder and Susan Webb Yackee, "Influence and the Administrative Process: Lobbying the U.S. President's Office of Management and Budget," *American Political Science Review* 109, no. 3 (August 2015): 507–522, https://www.thecre.com/forum2/wp-content/uploads/2015/08/haeder-and -yackee-2015-influence-and-the-administrative-process.pdf.

15. United States Environmental Protection Agency, "Standards of Performance for Greenhouse Gas Emissions from New, Modified, and Reconstructed Stationary Sources: Electric Utility Generating Units; Final Rule," *Federal Register* 80, no. 205 (October 23, 2015), https://www.govinfo.gov/content/pkg/FR-2015-10 -23/pdf/2015-22837.pdf.

16. Robin Bravender, "Obama Blindsides Enviros and EPA," *Politico*, September 2, 2011, https://www.politico.com/story/2011/09/obama-blindsides -enviros-and-epa-062586.

17. Kerwin and Furlong, *Rulemaking*.

18. Curtis W. Copeland, "Electronic Rulemaking in the Federal Government," Congressional Research Service, last updated May 16, 2008, https://fas.org/sgp /crs/misc/RL34210.pdf.

19. Sara Rinfret, Robert Duffy, and Jeffrey J. Cook, "Bots, Fake Comments, and E-Rulemaking: The Impact on Federal Regulations," *Public Administration Quarterly* (forthcoming).

20. Administrative Conference of the United States and Administrative Law Review, "Symposium: Mass and Fake Comments in Agency Rulemaking," October 18, 2018, https://www.acus.gov/transcript/mass-and-fake-comments-agency -rulemaking-transcript; Daniel A. Lyons, "E-Rulemaking and the Politicization of the Comment Process," *Digital Commons @ Boston College Law School*, October 23, 2018, https://lawdigitalcommons.bc.edu/cgi/viewcontent.cgi?article=2214& context=lsfp.

21. Nina A. Mendelson, "Rulemaking, Democracy, and Torrents of E-Mail," *George Washington Law Review* 79, no. 5 (2011): 1343–1380.

22. Jeffrey S. Lubbers, "Achieving Policymaking Consensus: The (Unfortunate) Waning of Negotiated Rulemaking," *South Texas Law Review* 49, no. 4 (2008): 987–1017.

23. Cary Coglianese, Heather Kilmartin, and Evan Mendelson, "Transparency and Public Participation in the Rulemaking Process: A Nonpartisan Presidential Transition Task Force Report," *George Washington Law Review* 77 (2009): 924.

24. Steven. J. Balla, Alexander R. Beck, William C. Cubbison, and Aryamala Prasad, "Where's the Spam? Interest Groups and Mass Comment Campaigns in Agency Rulemaking," *Policy and Internet* 11 (2019): 460–479, https:// onlinelibrary.wiley.com/doi/full/10.1002/poi3.224.

25. Kerwin and Furlong, *Rulemaking*.

26. Cynthia R. Farina, Mary J. Newhart, and Josiah Heidt, "Rulemaking v. Democracy: Judging and Nudging Public Participation That Counts," *Cornell e-Rulemaking Initiative Publications*, 2014, https://scholarship.law.cornell.edu

/cgi/viewcontent.cgi?referer=https://search.yahoo.com/&httpsredir=1&article
=1011&context=ceri.

27. Coglianese et al., "Transparency and Public Participation."

28. Rinfret et al., "Bots, Fake Comments, and E-Rulemaking."

29. Administrative Conference of the United States and Administrative Law
Review, "Symposium: Mass and Fake Comments in Agency Rulemaking."

30. Sheldon Kamieniecki, *Corporate America and Environmental Policy: How
Often Does Business Get Its Way?* (Stanford, CA: Stanford Law and Politics, 2006).

31. Jack Karsten and Darrell M. West, "Net Neutrality Debate Exposes
Weaknesses of Public Comment System," *Brookings*, January 18, 2018, https://
www.brookings.edu/blog/techtank/2018/01/18/net-neutrality-debate-exposes
-weaknesses-of-public-comment-system/.

32. James V. Grimaldi, "Government Watchdog Probes Fake Public Comments
on Regulations before Agencies," *Wall Street Journal*, January 24, 2018, https://
www.wsj.com/articles/government-watchdog-probes-fake-public-comments-on
-regulations-before-agencies-1516828844.

33. Marissa Martino Golden, "Interest Groups in the Rule-Making Process:
Who Participates? Whose Voices Get Heard?" *Journal of Public Administration
Research and Theory* 8, no. 2 (1998): 245–270; Cook, "Crossing the Influence Gap."

34. John M. Broder, "E.P.A. Issues Limits on Mercury Emissions," *New York
Times*, December 21, 2011, https://green.blogs.nytimes.com/2011/12/21/e-p-a
-announces-mercury-limits/.

35. Ibid.

36. Christopher McGrory Klyza and David J. Sousa, *American Environmental
Policy, 1990–2006 Beyond Gridlock* (Cambridge, MA: MIT Press, 2013).

37. Stuart Shapiro, "When Will They Listen? Public Comment and Highly
Salient Regulations," *Mercatus Institute*, September 5, 2013, https://www.mercatus
.org/publication/when-will-they-listen-public-comment-and-highly-salient
-regulations; Cook, "Crossing the Influence Gap."

38. Golden, "Interest Groups"; Cook, "Crossing the Influence Gap"; Kamieniecki,
Corporate America.

39. Michael E. Kraft and Sheldon Kamienicki, *Business and Environmental
Policy: Corporate Interests in the American Political System* (Cambridge, MA:
MIT Press, 2007); Kerwin and Furlong, *Rulemaking*.

40. Susan Webb Yackee, "Participant Voice in the Bureaucratic Policymaking
Process," *Journal of Public Administration Research and Theory* 25, no. 2 (2014):
427–449, doi:10.1093/jopart/muu007; Susan Webb Yackee, "Assessing Regulatory
Participation by Health Professionals: A Study of State Health Rulemaking," *Public
Administration Review* 73, no. 1 (2013): 105–114.

41. Shapiro, "When Will They Listen?"

42. Cook, "Crossing the Influence Gap"; Kamieniecki, *Corporate America*.

43. Jeffrey J. Cook and Sara R. Rinfret, "The EPA Regulates GHG Emissions:
Is Anyone Paying Attention?" *Review of Policy Research* 30, no. 3 (2013): 263–280.

44. Cook, "Crossing the Influence Gap."

45. Sara Rinfret and Michelle Pautz, *U.S. Environmental Policy in Action*, 2nd ed. (London: Palgrave Macmillan, 2019).

46. United States Environmental Protection Agency, "Prevention of Significant Deterioration and Title V Greenhouse Gas Tailoring Rule. Proposed Rule," *Federal Register* 74, no. 206 (2009): 55292–55365.

47. United States Environmental Protection Agency, "Prevention of Significant Deterioration and Title V GHG Tailoring Rule: EPA's Response to Public Comments," 2010, https://www.regulations.gov/document?D=EPA-HQ-OAR-2009-0517-19181.

48. United States Environmental Protection Agency and Department of Transportation, "2017 and Later Model Year Light-Duty Vehicle Greenhouse Gas Emissions and Corporate Average Fuel Economy Standards. Proposed Rule," *Federal Register* 76, no. 231 (December 1, 2011): 74854–75420, http://www.nhtsa.gov/staticfiles/rulemaking/pdf/cafe/2017-25_CAFE_NPRM.pdf.

49. United States Environmental Protection Agency and Department of Transportation, "2017 and Later Model Year Light-Duty Vehicle Greenhouse Gas Emissions and Corporate Average Fuel Economy Standards: EPA Response to Comments," 2012, https://nepis.epa.gov/Exe/ZyPURL.cgi?Dockey=P100EZXG.TXT.

50. Cook, "Setting the Record Straight."

51. Ibid.

52. United States Environmental Protection Agency and Department of Transportation, "2017 and Later . . . Response to Comments."

53. Cook, "Setting the Record Straight."

54. Jason Webb Yackee and Susan Webb Yackee, "A Bias Towards Business? Assessing Interest Group Influence on the U.S. Bureaucracy," *Journal of Politics* 68, no. 1 (2006): 128–139.

55. United States Environmental Protection Agency, "Prevention of Significant Deterioration."

56. Cook, "Setting the Record Straight."

57. Balla et al., "Where's the Spam?"

58. Cook, "Setting the Record Straight."

59. Helena Leino and Juha Peltomaa, "Situated Knowledge—Situated Legitimacy: Consequences of Citizen Participation in Local Environmental Governance," *Policy and Society* 31 (2012): 159–168.

60. Nasser Mohammed Al-Fannah, "Making Defeating CAPTCHAs Harder for Bots," *IEEE Computing Conference Proceedings*, 2017, https://ieeexplore.ieee.org/stamp/stamp.jsp?arnumber=8252183.

61. For more information on the University of Pennsylvania Program on Regulation, see https://www.pennreg.org/ and for Harvard's Regulatory Policy Program, see https://www.hks.harvard.edu/centers/mrcbg/programs/rpp/about.

62. Eisner et al., *Contemporary Regulatory Policy*.

Suggested Readings

Balla, Steven J., Alexander R. Beck, Elizabeth Meehan, and Aryamala Prasad. "Lost in the Flood? Agency Responsiveness to Mass Comment Campaigns in Administrative Rulemaking." *Regulation and Governance*, May 2020. doi:10.1111/rego.12318.

Kamienicki, Sheldon. *Corporate America and Environmental Policy: How Often Does Business Get Its Way?* Stanford, CA: Stanford Law and Politics, 2006.

Kraft, Michael E., and Sheldon Kamienicki. *Business and Environmental Policy: Corporate Interests in the American Political System.* Cambridge, MA: MIT Press, 2007.

Provost, Colin, and Brian Gerber. "The Contested Politics of Environmental Rulemaking." *Handbook of U.S. Environmental Policy.* Northampton, MA: Edward Elgar, 2020.

Bibliography

Administrative Conference of the United States and Administrative Law Review. "Symposium: Mass and Fake Comments in Agency Rulemaking." October 18, 2018. Available at https://www.acus.gov/transcript/mass-and-fake-comments -agency-rulemaking-transcript.

Al-Fannah, Nasser Mohammed. "Making Defeating CAPTCHAs Harder for Bots." In *Proceedings of the 2017 SAI Computing Conference.* New York: IEEE, 2017. Available at https://ieeexplore.ieee.org/stamp/stamp.jsp?arnumber=8252183.

Balla, Steven J., Alexander R. Beck, William C. Cubbison, and Aryamala Prasad, "Where's the Spam? Interest Groups and Mass Comment Campaigns in Agency Rulemaking." *Policy and Internet* 11 (2019): 460–479. Available at https://onlinelibrary.wiley.com/doi/full/10.1002/poi3.224.

Bravender, Robin. "Obama Blindsides Enviros and EPA." *Politico*, September 2, 2011. Available at https://www.politico.com/story/2011/09/obama-blind sides-enviros-and-epa-062586.

Broder, John M. "E.P.A. Issues Limits on Mercury Emissions." *New York Times*, December 21, 2011. Available at https://green.blogs.nytimes.com/2011/12/21 /e-p-a-announces-mercury-limits/.

Coglianese, Cary, Heather Kilmartin, and Evan Mendelson. "Transparency and Public Participation in the Rulemaking Process: A Nonpartisan Presidential Transition Task Force Report." *George Washington Law Review* 77 (2009): 924.

Collins, Keith. "Net Neutrality Has Officially Been Repealed. Here's How That Could Affect You." *New York Times*, June 11, 2018. Available at https://www .nytimes.com/2018/06/11/technology/net-neutrality-repeal.html.

Cook, Jeffrey J. "Crossing the Influence Gap: Clarifying the Benefits of Earlier Interest Group Involvement in Shaping Regulatory Policy." *Public Administration Quarterly* 42, no. 4 (2018).

———. "Setting the Record Straight: Interest Group Influence on Environmental Protection Agency Climate Policy." Unpublished diss., Colorado State University, 2017. Available at https://dspace.library.colostate.edu/bitstream /handle/10217/181399/Cook_colostate_0053A_14127.pdf?sequence=1.

Cook, Jeffrey J., and Sara R. Rinfret. "The EPA Regulates GHG Emissions: Is Anyone Paying Attention?" *Review of Policy Research* 30, no. 3 (2013): 263–280.

Copeland, Curtis W. "Electronic Rulemaking in the Federal Government." Congressional Research Service. Last updated May 16, 2008. Available at https:// fas.org/sgp/crs/misc/RL34210.pdf.

Eisner, Mark Allen, Jeffrey Worsham, and Evan J. Ringquist. *Contemporary Regulatory Policy.* London: Lynne Rienner, 2006.

Farina, Cynthia R., Mary J. Newhart, and Josiah Heidt. "Rulemaking v. Democracy: Judging and Nudging Public Participation That Counts." *Cornell e-Rulemaking Initiative Publications*, 2014. Available at https://scholarship .law.cornell.edu/cgi/viewcontent.cgi?referer=https://search.yahoo.com/& httpsredir=1&article=1011&context=ceri.

Fleishman, Glenn. "FCC Chair Ajit Pai Admits Millions of Russian and Fake Comments Distorted Net Neutrality Repeal." *Fortune*, December 5, 2018. Available at http://fortune.com/2018/12/05/fcc-fraud-comments-chair-admits/.

Golden, Marissa Martino. "Interest Groups in the Rule-Making Process: Who Participates? Whose Voices Get Heard?" *Journal of Public Administration Research and Theory* 8, no. 2 (1998): 245–270.

Grimaldi, James V. "Government Watchdog Probes Fake Public Comments on Regulations Before Agencies." *Wall Street Journal*, January 24, 2018. Available at https://www.wsj.com/articles/government-watchdog-probes-fake-public -comments-on-regulations-before-agencies-1516828844.

Haeder, Simon F., and Susan Webb Yackee. "Influence and the Administrative Process: Lobbying the U.S. President's Office of Management and Budget." *American Political Science Review* 109, no. 3 (August 2015): 507–522. Available at https://www.thecre.com/forum2/wp-content/uploads/2015/08/haeder -and-yackee-2015-influence-and-the-administrative-process.pdf.

Johnson, Ted. "FCC Inspector General: John Oliver Segment Triggered System Slowdown, Not Bots." *Variety*, August 7, 2018. Available at https://variety .com/2018/politics/news/john-oliver-fcc-triggered-slowdown-1202898328/.

Kamieniecki, Sheldon. *Corporate America and Environmental Policy: How Often Does Business Get Its Way?* Stanford, CA: Stanford Law and Politics, 2006.

Karsten, Jack, and Darrell M. West. "Net Neutrality Debate Exposes Weaknesses of Public Comment System." *Brookings*, January 18, 2018. Available at https:// www.brookings.edu/blog/techtank/2018/01/18/net-neutrality-debate-exposes -weaknesses-of-public-comment-system/.

Kerwin, Cornelius M., and Scott Furlong. *Rulemaking: How Government Agencies Write Law and Make Policy.* 3rd ed. Washington, DC: CQ Press, 2018.

Klyza, Christopher McGrory, and David J. Sousa. *American Environmental Policy, 1990–2006 Beyond Gridlock.* Cambridge, MA: MIT Press, 2013.

Kraft, Michael E., and Sheldon Kamienicki. *Business and Environmental Policy: Corporate Interests in the American Political System*. Cambridge, MA: MIT Press, 2007.

Leino, Helena, and Juha Peltomaa. "Situated Knowledge—Situated Legitimacy: Consequences of Citizen Participation in Local Environmental Governance." *Policy and Society* 31 (2012): 159–168.

Locker, Melissa. "John Oliver Wants You to Flood the FCC Website to Save Net Neutrality, Again." *Time*, May 8, 2017. Available at https://time.com/4770205/john-oliver-fcc-net-neutrality/.

Lubbers, Jeffrey S. "Achieving Policymaking Consensus: The (Unfortunate) Waning of Negotiated Rulemaking." *South Texas Law Review* 49, no. 4 (2008): 987–1017.

Lyons, Daniel A. "E-Rulemaking and the Politicization of the Comment Process." *Digital Commons @ Boston College Law School*, October 23, 2018. Available at https://lawdigitalcommons.bc.edu/cgi/viewcontent.cgi?article=2214&context=lsfp.

Mendelson, Nina A. "Rulemaking, Democracy, and Torrents of E-Mail." *George Washington Law Review* 79, no. 5 (2011): 1343–1380.

Rinfret, Sara, Robert Duffy, and Jeffrey J. Cook. "Bots, Fake Comments, and E-Rulemaking: The Impact on Federal Regulations." *Public Administration Quarterly* (forthcoming).

Rinfret, Sara, and Michelle Pautz. *U.S. Environmental Policy in Action*. 2nd ed. London: Palgrave Macmillan, 2019.

Shapiro, Stuart. "When Will They Listen? Public Comment and Highly Salient Regulations." *Mercatus Institute*. September 5, 2013. Available at https://www.mercatus.org/publication/when-will-they-listen-public-comment-and-highly-salient-regulations.

United States Environmental Protection Agency. "Fact Sheet: Clean Power Plan Key Changes and Improvements: From Proposal to Final." Last modified May 9, 2017, Available at https://archive.epa.gov/epa/cleanpowerplan/fact-sheet-clean-power-plan-key-changes-and-improvements.html.

———. "Prevention of Significant Deterioration and Title V GHG Tailoring Rule: EPA's Response to Public Comments." 2010. Available at https://www.regulations.gov/document?D=EPA-HQ-OAR-2009-0517-19181.

———. "Prevention of Significant Deterioration and Title V Greenhouse Gas Tailoring Rule. Proposed Rule." *Federal Register* 74, no. 206 (2009): 55292–55365.

———. "The Safer Affordable Fuel Efficient (SAFE) Vehicles Proposed Rule for Model Years 2021–2026." September 27, 2018. Available at https://www.epa.gov/regulations-emissions-vehicles-and-engines/safer-affordable-fuel-efficient-safe-vehicles-proposed.

———. "Standards of Performance for Greenhouse Gas Emissions from New, Modified, and Reconstructed Stationary Sources: Electric Utility Generating Units; Final Rule." *Federal Register* 80, no. 205 (October 23, 2015). Available at https://www.govinfo.gov/content/pkg/FR-2015-10-23/pdf/2015-22837 .pdf.

United States Environmental Protection Agency and Department of Transportation. "2017 and Later Model Year Light-Duty Vehicle Greenhouse Gas Emissions and Corporate Average Fuel Economy Standards: EPA Response to Comments." 2012. Available at https://nepis.epa.gov/Exe/ZyPURL.cgi ?Dockey=P100EZXG.TXT.

———. "2017 and Later Model Year Light-Duty Vehicle Greenhouse Gas Emissions and Corporate Average Fuel Economy Standards. Proposed Rule." *Federal Register* 76, no. 231 (December 1, 2011): 74854–75420. Available at http://www .nhtsa.gov/staticfiles/rulemaking/pdf/cafe/2017-25_CAFE_NPRM.pdf.

Wagner, Wendy, Katherine Barnes, and Lisa Peters. "Rulemaking in the Shade: An Empirical Study of EPA's Air Toxic Regulations." *Administrative Law Review* 63, no. 1 (2011): 99–158.

Yackee, Jason Webb, and Susan Webb Yackee. "A Bias Towards Business? Assessing Interest Group Influence on the U.S. Bureaucracy." *Journal of Politics* 68, no. 1 (2006): 128–139.

Yackee, Susan Webb. "Assessing Regulatory Participation by Health Professionals: A Study of State Health Rulemaking." *Public Administration Review* 73, no. 1 (2013): 105–114.

———. "Participant Voice in the Bureaucratic Policymaking Process." *Journal of Public Administration Research and Theory* 25, no. 2 (2014): 427–449. Available at https://doi.org/10.1093/jopart/muu007.

4

Is the Rule Final Yet?

Executive Orders, Litigation,
and Rule Finalization

Deserai A. Crow, Lydia A. Lawhon, and Sara K. Guenther

The warty, three-inch-long dusky gopher frog may seem an un-likely character in the regulatory process. However, given its penchant for living on private timberland in Louisiana and its status as a federally endangered species, the frog became the focal point of a protracted legal battle that was decided in *Weyerhaeuser v. US Fish and Wildlife Service* (No. 17-71), a 2018 Supreme Court case. Weyerhaeuser, a prominent timber company, argued against the decision of the U.S. Fish and Wildlife Service to designate un-occupied private land as critical habitat, citing economic concerns.[1] The Supreme Court agreed. At issue was the definition of *habitat*, which remained unclarified even in the court's decision. Although the Trump administration refined the definition of *critical habitat* in new rules promulgated in August 2019, this broader habitat ques-tion remained unsettled: "We note that we do not in the rule attempt to definitively resolve the full meaning of the term 'habitat.'"[2] In late December 2019, the Trump administration proposed tackling this challenge through another round of rulemaking.

We usually think of rulemaking as a linear or iterative process. Rule writers issue a draft rule and then solicit and examine public comment, as explained in Chapter 1. The agency considers this feedback and subsequently publishes a final rule in the *Federal Register.* Yet, in environmental rulemaking, rule finalization can be riddled with controversy. Competition among political institutions to influence the final rule is common.

This chapter explores the key institutional actors and their impact on the finalization of U.S. environmental rules, specifically describing how Congress, the courts, and the executive office all attempt to influence rule finalization after the formal rulemaking process has ended. A case study of examples of rulemaking involving the Endangered Species Act (ESA) then highlights post-rulemaking battles as they play out among federal institutions. The Endangered Species Act provides an excellent case study of the use of the primary tools of authority for presidents and Congress—executive orders, political appointments, rollbacks through legislation, and riders to appropriations bills—as well as examples of how political actors use the court system to mitigate conflicts over rulemaking activity. The ESA provides numerous examples where federal institutions—Congress, the courts, and the executive office—jockey for influence over implementation of the law using the various tools as their disposal.

The Presidency and the Final Rule

In this section, we discuss how the executive branch, through tools such as political appointments and executive orders, influences rulemaking activity. Rules originating from the executive branch have the support of the president, but only for rules promulgated by the president's administration. A president delegates authority to an agency to create regulations that comport with the administration's priorities. When a president disagrees with a rule inherited from a prior administration or one from the agency's past actions, the executive branch typically orders new rules to loosen or strengthen regulations based on the political preferences of the new administration. New administrations can utilize a number of tactics to thwart "midnight" rules—or those made at the end of the previous administration. These tactics include directing that any final or proposed rules not be sent to the Office of the Federal Register until they have been approved by the new administration, or issuing a sixty-day postponement on the effective date for new regulations.

For example, just before the end of his term, President George W. Bush's administration issued a rule allowing federal agencies to bypass a consultation process mandated by the Endangered Species Act. The consultation process ensured that government activities and development projects did not interfere with endangered species or their habitats. The new Bush rule intended to reduce the regulatory burden of the ESA in favor of industry. Within the first few months of his presidency, however, President Barack Obama signed a memorandum suspending the Bush administration's rule change and restoring the former guidelines of the consultation process. In the period after the 2016 election and before the inauguration of President Trump in January 2017, the Obama administration utilized the same tactic as the Bush administration and passed several rules at the last minute in anticipation of a new executive driven by an entirely different ideology.

As noted in Chapter 3, considerable evidence exists that insiders to the rulemaking process—those who have routinely engaged in rulemaking proceedings and are often from the regulated industries—hold the most influence during rulemaking proceedings. Other stakeholders often participate in public comment and attempt to influence rulemaking decisions, but for a variety of reasons, including technical knowledge, resources, prior rulemaking experience, and agency capture, they are often less successful.[3]

Political Appointments and Direct Presidential Influence on Rulemaking

Through presidential appointments to agency leadership and management positions, presidents powerfully signal their political and policy priorities. Political appointees help carry out a president's priorities through budgeting, management prioritization, programmatic modifications, and messaging tools.[4] These political actors can significantly change the direction of an agency, as James Watt did with the Environmental Protection Agency (EPA) while serving as secretary of the interior under the Reagan administration, reducing the regulatory burden on industry and the stringency of environmental protections.[5] Changes in agency behavior are particularly prominent with changes in party control in the White House and can include major organizational restructuring under the leadership of presidential appointees.[6] For example, under the Trump administration, both the secretary of agriculture and the interior secretary oversaw the relocation of agencies' headquarters from the District of Columbia to Midwestern and western cities, which critics feared would give local economic interests, rather than national interests, leverage over federal rulemaking and raised concerns about diminished capacity within agencies as many longtime employees elected to quit or seek early retirement rather than move.[7]

Since presidential appointment power can be so consequential, other branches of government do have opportunities to weigh in. Presidential nominations for most top agency and management positions require Senate confirmation, an opportunity for the Senate to check presidential power. The courts can also limit the president's political influence over rulemaking and policy implementation, but very little judicial precedence or clarity exists on the matter.[8] Since the 1960s, the number of appointed positions in the executive branch has nearly doubled, in part because presidents seek to control rulemaking.[9] Both Democratic and Republican presidents have tested and expanded the influence of their powers over rulemaking, and courts have yet to constrain them.

Executive Orders

One of the more powerful tools presidents have in their tool kit is the issuance of executive orders. President Reagan's Executive Order 12291, for example, decreed that presidential appointees to federal executive agencies would have access to rulemaking proceedings and could determine whether a rule was a "major" change, thereby making it subject to cost-benefit analysis requirements.[10] Immediately, this executive order was subject to legal review due to its far-reaching nature and potential infringement on the legislative role granted to Congress.[11] While this executive order codified trends that had been observed in prior presidential administrations, the effect on agency rulemaking was felt acutely by agencies such as the EPA and others, where cost-benefit analysis of regulations to address diseases, threats to public health, and other societal problems can be far more difficult to quantify than rules in other realms.[12] The approach formalized under the Reagan administration continued through the H. W. Bush presidency.

Presidents Clinton and W. Bush continued using this regulatory cost-effectiveness guideline as part of the toolbox of presidential influence over rulemaking.[13]

Under the Trump administration, a renewed emphasis was placed on dismantling federal regulatory institutions—the agencies themselves—and the regulations that govern environmental protection in the United States. Early in the Trump presidency, the administration's priorities were focused on rolling back rules promulgated toward the end of the Obama administration. Focus then shifted toward changing the strength of many of the nation's most broad and powerful laws. For example, the redefinition of critical habitat discussed in this chapter, may have lasting impacts on the ability of the federal government to protect species and their habitats. The Trump administration also set its sights on changing the National Environmental Policy Act's requirements for public review of major infrastructure projects. Critics argue that this simple change weakens one of the nation's most powerful environmental protection tools.[14] By focusing on altering the broad tools for environmental protection, these regulatory changes may have lasting effects on environmental policy and management in the United States for many years to come.

Through the power to manage the federal bureaucracy, to appoint leadership and management positions throughout government, and to use executive orders to make policy and programmatic changes, presidents have routinely tried to expand their influence over rulemaking, as is highlighted in the ESA case section later in this chapter. Although ample debate takes place over the legality of these expansions, to date the courts have been unwilling to place a check on this type of presidential power when Congress leaves laws open to interpretation and agency rulemaking authority.

Congress's Powers to Override
or Undo Rulemaking Decisions

As noted in Chapter 1, Congress defines an agency's scope of authority when it creates a new agency, but executive agencies retain some degree of discretion to combine legislative record and statutory elements to formulate an acceptable rule.[15] Within this typical range of discretion, politics—not law—prevails, which can create conflict among the legislative and executive branches and lead to institutional wrestling over the final rule outcomes.[16] The concept of legislative supremacy is core to the U.S. system of governance, as it is to other constitutional democracies as well.[17] Scholars view legislatures—either the Congress or state legislatures—as the appropriate and primary locus of policy decision making, arguing that part of the framers' intent was to create Congress as the dominant branch, which is why they gave the legislature near exclusive control over the budget.[18] Budgets for government agencies as well as specific spending instructions or prohibitions for implementing laws are part of the typical scope of legislative authority, meaning that even existing laws can be affected based on budget allocations for implementation.[19]

Beyond legislative power to create laws and allocate funding for government activities, Congress also has tools to roll back or contravene executive branch rulemaking decisions. Since the mid-1970s, Congress has passed hundreds of overrides, or laws that explicitly seek to reverse or modify judicial interpretations of statutes.[20] In addition to legislative overrides, Congress gave itself the power to review all federal regulations in 1996 when it passed the Congressional Review Act (CRA). The CRA removes the courts from the substantive review process for agency lawmaking and allows Congress to insert itself in rulemaking decisions with which it disagrees or which it believes conflict with the intent of a law.[21] It established congressional review of administrative rules, which requires agencies to submit rules to Congress for review. As a result, federal agency rules are no longer final on the day the agency announces as their effective date.[22] Congress has sixty days to overturn a rule following its publication. It is important to note the CRA has not often been used in practice, with the exception of a period after Trump took office and Congress overturned a number of Obama administration rules.

Congressional review of agency rules can delay new regulations for several months. This oversight role for Congress was originally thought to have a chilling effect on the bureaucracy that no longer had the autonomy it had previously enjoyed before the CRA was passed in 1996.[23] Congress enacted, and President George W. Bush signed into law, the first joint resolution of this nature in March 2001, disapproving an ergonomics standard issued by the Occupational Safety and Health Administration. In expressing support for this action, members of Congress and President Bush stated that their purpose in using the CRA to disapprove this rule was to increase accountability for agency rulemaking.[24] If a rule is disapproved using the CRA, the agency may not reissue the same or a substantially similar rule unless the agency has been provided specific statutory authority enacted after the date of the joint resolution disapproving the original rule.[25]

The Courts as a Venue of Recourse

Political actors often turn to the courts when rulemaking does not align with their preferred policy outcomes or interpretations of existing law and when they do not have allies to pursue recourse in the executive or legislative branches. The term *venue shopping* refers to the choices interest group advocates make regarding which policy-making institutions will be on the receiving end of their advocacy as they strive to mobilize political support for their members' interests.[26] Congressional committees, state government organizations, courts, and private businesses are potential venues of advocacy action.[27] Venue shopping is common among policy actors who are not among the dominant participants in an arena and who want to force policy change. Advocates who are unable to make headway at the state level may refocus their efforts locally, for example.[28] When there are many venues available, interest groups usually target venues where they have exchange relationships—where they hold influence due to the reciprocal nature of the relationship between advocacy groups and the venue (legislature, courts, agencies, etc.).[29]

In environmental law and rulemaking, advocates for environmental protection engage strategically in venue shopping in the courts. For a party to intervene in a governmental dispute via the courts, the party must prove it has standing to sue. The traditional model of prudential standing establishes that legal standing is based on (1) the prohibition against litigating generalized grievances, (2) the prohibition against litigating the rights of a third party, and (3) the requirement that the plaintiff's interest falls within the zone of interests the statute was designed to protect.[30] Congress can expand who has standing to sue, which is referred to as establishing statutory standing.[31] Environmental case law has established that many environmental laws allow for citizens to file suit against the government for failing to enforce a law or against a third party for environmental harm. The requirement of a tangible, particularized injury to the plaintiff was required under cases such as *Lujan*, but more recent case law in *Laidlaw* allows for more speculative claims of perceived threats to a statutorily recognized interest from governmental action and can establish a "sufficient stake" in the matter to prove standing.[32]

What Role for the States?

The federal government plays a dominant role in policy making in the United States through the executive branch, Congress, and the court system, but states can play a central role in creating their own laws as well. States partner with federal agencies to enforce federal laws with varying degrees of stringency, and sometimes go beyond federal benchmarks. Federal policy decisions should have some effect on state policy choices concerning the existence, scope, and contours of state regulatory programs.[33]

States have two main roles with regard to federal rulemaking. First, states play a prominent and influential role in the monitoring and implementation of some laws. This role is key to understanding how laws such as the Clean Air Act work.[34] With incentives from the federal government and the threat of sanctions for noncompliance, states monitor and enforce the Clean Air Act's stationary sources of air pollution. Second, in environmental regulation, the federal government often provides states the ability to promulgate a rule that is state-specific in order to consider their local resources, industries, and needs. One such example is the issuance of effluent emissions standards for concentrated animal feeding operations. The EPA issued a rule requiring all U.S. states to set a standard that was responsive to local water table conditions, hydrology, and industry (hogs and cattle entail different requirements, for example).[35]

Finally, recent trends in environmental policy and management indicate that collaboration is now encouraged more than in the past, when conflict among stakeholders—whether state/federal or industry/environmental advocacy—was ever-present. As collaborative forms of governance become more common across the country, as observed in water management in the western United States,[36] this trend toward collaboration and coordination may show increasing success.[37] This form of environmental management and policy making creates ample room for state involvement and perhaps more room for states to influence environmental policies and outcomes. In some cases, states may be able to preempt federal intervention when collaboration with stakeholders is done proactively and effectively.

Executive branch agencies wield considerable power through their roles in implementing and enforcing laws. The role of bureaucrats in shaping implementation is central to the conversation of public administration and bureaucracy scholars.[38] Street-level bureaucrats—the people who interact with citizens during their implementation of laws and who are tasked with the day-to-day operations of government—can play a powerful role in interpreting laws and translating those laws into action and outcomes.[39]

Congress's usual approach to enlist states in the implementation and enforcement of federal regulatory programs is to offer technical and financial assistance to state agencies and to threaten to cut off federal public works funding or to increase regulatory burdens on industries in uncooperative states. Normally Congress is able to induce states to cooperate in implementing and enforcing federal environmental policies.[40] Past studies have shown that environmental enforcement by the federal government is influenced more by expectations of state compliance—usually related to political factors and expectations of state-level cooperation with environmental rules—than evidence of violations.[41]

Environmental inspectors are included in this category (see Chapters 1 and 5). They monitor regulated facilities to ensure the facilities are in compliance with environmental regulations.[42] The inspector serves as a key liaison between government and regulated facilities. Therefore, a firm's experiences with government (at least as they relate to environmental protection) are largely the result of the interactions its owner and employees have with the inspector.[43] Like other street-level bureaucrats who translate and deliver wider standards set by others (e.g., teachers, social workers, police officers), what inspectors do effectively becomes agency policy, and the nature and extent of environmental protection is formed by this process.[44]

The importance of inspectors cannot be overstated. Economists and policy makers generally believe that effective pollution regulations require recurrent inspections and sanctions. The enforceability of regulations is a dominant factor in dramatic improvements in developed countries' environmental quality since the 1980s.[45]

Case Study: The Endangered Species Act

As noted at the beginning of this chapter, the Endangered Species Act provides a useful lens through which to explore the tools of the executive office and Congress, as well as the roles of federal court system and the states, in the context of rule finalization.

Congress passed the ESA in 1973 at the urging of President Nixon in his State of the Union address, and with near-unanimous consent of Congress. The law strengthened the existing federal endangered species protection program by: (1) empowering the U.S. Fish and Wildlife Service (USFWS) to list species at risk of extinction as either endangered or threatened, based on best available scientific data;[46] (2) allowing private citizens to petition species for listing, and the ability to sue to enforce the law; (3) prohibiting federal agencies, or any project aided by federal money, to harm listed species or their habitats; and (4) making the act of harming a listed species or its critical habitat a federal offense.

The ESA included provisions granting authority to the USFWS and the National Marine Fisheries Service (NMFS) (now NOAA [National Oceanic and Atmospheric Administration] Fisheries) to implement and enforce the ESA; these provisions do not expire. It also included a provision authorizing appropriations, which were set to expire at the end of fiscal year 1978. The reauthorization of the appropriations process set the ESA on a schedule for consideration and possible amendment by Congress. In other words, reauthorization gave Congress an opportunity to evaluate implementation of the ESA by the executive and judicial branches and amend the statutory language accordingly. Reauthorizations of the ESA were passed in 1978, 1982, and 1988, and each reauthorization included amendments to various sections of the law; these are discussed in more detail in the section on congressional attempts to roll back the ESA.

Sections four and seven of the ESA have been the focus of rulemaking activity. Section four identifies species eligible for listing and the factors that need to be considered in determining listing decisions, provides guidance in creating recovery plans, and addresses critical habitat considerations for protecting threatened and endangered species.[47] Section seven of the ESA mandates that agencies engage in consultations with either the USFWS or NMFS on any actions they undertake that may jeopardize protected species. This section is considered one of the primary tools for protecting species by advocates of the ESA, while it is conversely critiqued as a liability for economic development due to its cost and impediment of the time line of various projects and is often the target of rulemaking revisions.[48] For example, the Revision of Regulations for Interagency Cooperation rule,[49] promulgated in 2019, intends to address the timeliness of ESA consultations by setting deadlines for informal consultations and other measures to streamline the consultation process.[50]

Congressional Attempts at ESA Rollback

Congress has made numerous attempts via legislation to change the statutory language mandating agency implementation and enforcement activities related to the ESA. The Center for Biological Diversity collects data on these legislative attempts to weaken the ESA, as presented in Figure 4.1. These initiatives include both species-specific and general proposals to amend the ESA. Species-specific proposals attempt to permanently delist, temporarily delist, or delay listing of particular species and are introduced by members of Congress representing the states or districts inhabited by the targeted species. The most common species targeted are the gray wolf, the delta smelt, the greater sage-grouse, the lesser prairie chicken, the northern long-eared bat, and the American burying beetle. General proposals vary from placing constraints on USFWS authority to list new species—including eliminating funding to list new species, designating critical habitat, and monitoring existing endangered species—to requiring the federal government to compensate property owners affected by the ESA or prohibiting the consideration of climate change in decisions to list new species.

Only sixteen out of 445 bills in the Center for Biological Diversity data set have been signed into law, with the majority of these being riders to appropriations bills. For example, Congress was successful in using the budget process to delist gray wolves in Montana and

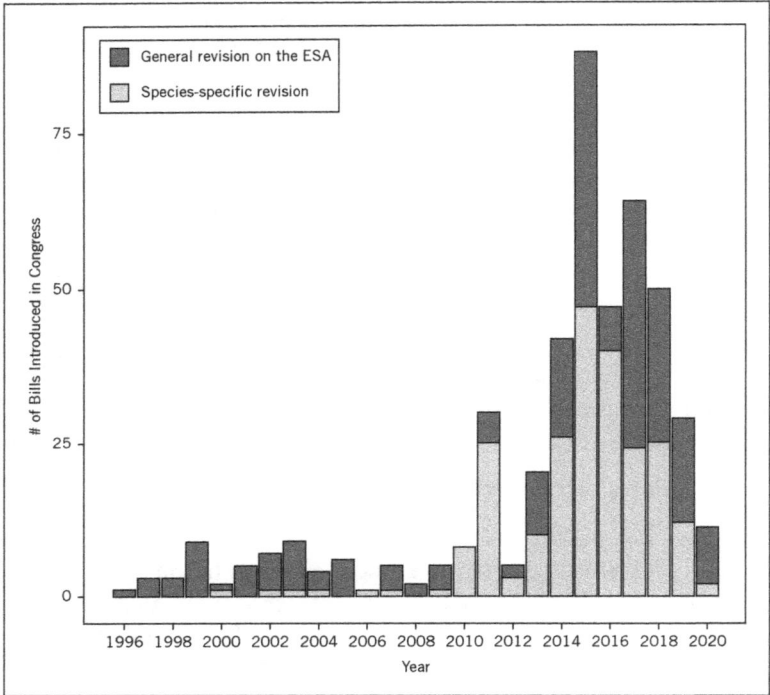

Figure 4.1   Number of Bills Introduced in Congress That Attempt to
Weaken the Endangered Species Act, 1996–2019
(data compiled by the Center for Biological Diversity,
https://www.biologicaldiversity.org/campaigns/esa_attacks/table.html)

Idaho in 2011 via a budget rider in the Department of Defense and
Full-Year Continuing Appropriations Act. The remaining proposals
have either failed in the House or Senate (68 percent), have been vetoed by the president (3 percent), or are still under consideration as
of this writing (25 percent).

The President, Congress, and the Endangered Species Act

The ESA has endured persistent tousling for authority among the
branches of government. This section describes how congressional
and presidential interests drive success or failure of ESA revision.

Note that the ESA's authorization of appropriation expired in 1992. In fact, the ESA has yet to be reauthorized as of early 2021 Though legislators have been unable to reach consensus on a revision of the ESA, Congress still annually appropriates funding for ESA-related programs. The reason Congress continues to fund the ESA is that administrative action is required to carry out ESA mandates, including consultations and the processing of permits. The absence of this administrative effort would force development projects to stall or subject them to litigation, frustrating development interests and potentially harming the economy. The following paragraphs describe the political context of successful and unsuccessful attempts to revise the ESA under each presidential administration.

Carter and the 1978 Amendments

The power of the ESA as written became clear within just a few years of the law's enactment. A small fish, the snail darter, was listed as endangered and legally prevented the completion of a dam in the Tennessee River Valley, a federal project that had begun before the ESA's passage. Though a majority of public works projects carried on just fine with modifications to accommodate endangered species, members of Congress worried that continued enforcement of the ESA would threaten "pet public-works projects in their districts."[51] Congress made the first amendment to the ESA allowing exemptions by way of the "God Squad," a Cabinet-level committee that could vote to exempt projects from the ESA even if they caused a species to go extinct.[52]

Reagan and the 1982 Amendments

In 1982, the ESA was due for reauthorization. It was an opportunity for Congress to put pressure on the new Reagan administration, which had taken significant actions to favor industry in policy implementation. President Reagan issued an executive order requiring all federal agencies to estimate the economic costs of their actions. Furthermore, the administration slowed to a near halt the process of listing new species as endangered or threatened. Congress, led by a Democratic majority in the House and a Republican majority in the Senate, responded to the administration by shortening deadlines for the USFWS, forcing the agency to respond to petitions for listing species more quickly, and explicitly prohibiting the USFWS from considering economic costs in its listing decisions.

Congress added a few important amendments to the ESA in the 1982 reauthorization that offered new opportunities for exemption from the law. One exemption granted the secretary of the interior the authority to classify "experimental populations" within an endangered species and relax restrictions associated with those populations under section 10(j) of the ESA. To designate a species as an experimental population, the USFWS must oversee the reintroduction of the species, and the area for reintroduction must be geographically isolated from other individuals of that species.[53] Prior to this exemption, the USFWS was empowered to reintroduce endangered species to areas of that species' historic range in order to establish a new population as part of recovery efforts. However, reintroduction proposals often elicited strong local opposition, primarily based on concerns that these actions would result in land use restrictions on both federal and nonfederal land. The exemption for experimental populations was added to the ESA to foster cooperation from local communities and increase the likelihood of successful reintroduction efforts.[54] The 10(j) rule was instrumental in facilitating the reintroduction of wolves to Yellowstone and central Idaho in 1995 and 1996, because it provided local communities reassurance that the USFWS would have more flexibility in management options, such as lethal control for animals that caused livestock depredations.

Another type of exemption introduced in the 1982 amendments permitted incidental taking of endangered species. *Taking* is defined as harming an individual of an endangered species or the habitat necessary for its survival. Each application for an incidental take permit was to be accompanied by a Habitat Conservation Plan, which would outline measures taken by the applicant to reduce or mitigate any harm to the species covered in the plan. The CQ Almanac called the 1982 ESA reauthorization "one of the year's more significant accomplishments."[55]

H. W. Bush and Congress on Standby

The 1990 listing of the northern spotted owl as threatened provided the legal framework for environmentalists to use the ESA to protect old-growth forest, the owl's preferred habitat. The northern spotted owl crisis exacerbated partisan and regional divisions in Congress, limiting any chance at consensus for reauthorization. In 1989 and 1991, federal judge William Dwyer twice ordered logging in the Pacific Northwest to stop on the grounds that federal agencies were not doing enough to protect the northern spotted owl under the ESA. These court orders triggered the Bureau of Land Management to file for an exemption with the God Squad, which eventually voted 5–2 to exempt logging on thirteen of the forty-four tracts from restrictions under the ESA, though this decision was later overturned in court.[56]

As the 1992 presidential election approached and ESA appropriation authorization was set to expire, congressional party leaders were reluctant to bring up the divisive issue and risk alienating interest groups by coming down on one side or the other—or worse, alienating both sides. The presidential candidates campaigned on the issue, with President H. W. Bush vowing not to sign a reauthorization of the ESA that did not take into account the economic costs of protecting wildlife, making ESA reauthorization untenable.

Clinton and the Republican Revolution

During the first two years of the Clinton administration, great effort and much attention were placed on brokering a deal between timber companies and environmentalists over the northern spotted owl and old-growth forest management, including a conference led by the president himself. This conflict bled into the broader discussion of ESA reauthorization, and in 1993, bills that included concessions to property rights and states' interests and concessions to environmentalists were proposed in the House.[57] One ESA reauthorization bill required the federal government to compensate property owners who experience loss in property value due to the ESA. ESA reauthorization proposals had shifted toward deregulation after the Republican Party gained control of the House and Senate in 1995. In addition to ESA overhaul attempts, which largely failed, Republicans used budget riders to chip away at ESA provisions they opposed. At this point, the ESA issue was so polarizing that even a moderate and bipartisan reauthorization bill reported out of committee in 1997 did not satisfy a sufficient number of senators on either side of the issue to bring it to a vote.

George W. Bush and Gridlock in the Senate

Congressional efforts during the first George W. Bush administration were part of a Republican strategy to "work at the margins of the law, proposing limited changes that will have bipartisan backing" as opposed to proposing significant changes to the ESA that were unlikely to pass in the Senate.[58] Despite a mostly unified government during the second Bush administration, the Senate was just moderate enough to stall any legislative attempts at ESA reform originating from the more conservative House. Republican moderates and Democrats, on the other hand, dared not risk introducing a moderate reauthorization bill out of fear it would be hijacked by conservatives.[59]

Obama and Attempts at "Piecemeal" ESA Reform

Reauthorization of the ESA was not a priority for Democrats during their period of unified government between 2009 and 2011, when their primary focus was health care reform. Beginning in 2011, however, when Republicans regained majority party status in the House, attempts to scale back the ESA resumed. The first line of attack came through the appropriations process. House Republicans added a rider to the Department of the Interior appropriations bill that would prohibit the Fish and Wildlife Service from using federal funds toward listing new species or increasing protections for already listed species.[60] Obama threatened to veto the bill, and Democrats—with the help of thirty-seven Republicans—were successful in removing the moratorium from the spending bill when it was brought to the floor.

The Republicans' piecemeal strategy for ESA reform was not limited to appropriations bills. Legislators also debated ESA-related amendments in the reauthorization of the farm bill in 2013 and in the water infrastructure bill in 2016, to name a few. In a letter to Obama the following year, ninety-two House Democrats urged the president to "maintain steadfast opposition" to any attempts to undermine the ESA, referencing "damaging anti-wildlife provisions" in the Interior appropriations bill, the Department of Defense reauthorization bill, and House-passed energy package.

This pressure from Congress to roll back provisions in the ESA may have influenced the Obama administration to streamline implementation of the ESA through an increased reliance on special rules under section 4(d) (see Figure 4.2). These rules exempt certain

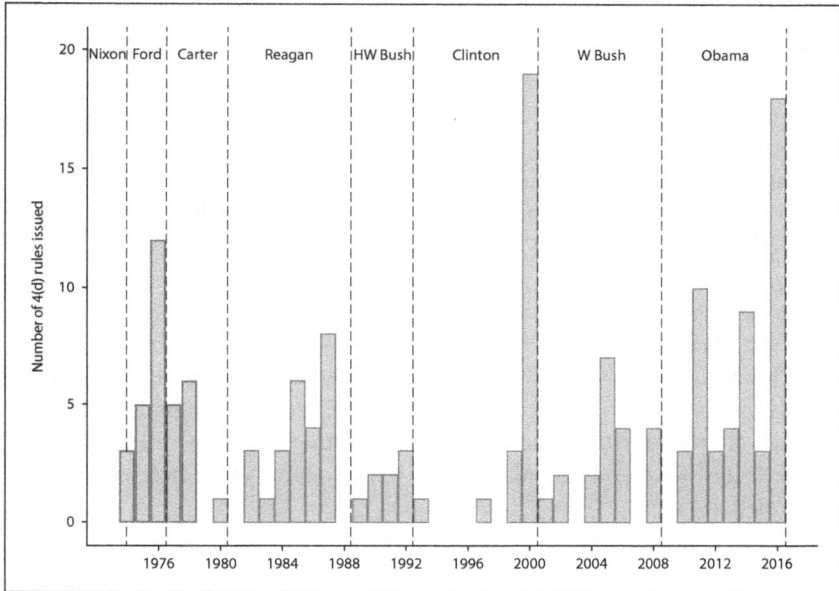

Figure 4.2 Number of Endangered Species Act Section 4(d) Rules Issued, 1976–2016 (data from Ya-Wei Li, "Section 4(d) Rules: The Peril and the Promise," Defenders of Wildlife ESA Policy White Paper Series, 2017)

activities from needing incidental take permits from the USFWS or NMFS in order to proceed. Depending on the activities these rules authorize—which range from conservation and research activities to timber harvest, fishing, and livestock maintenance—the concern is that overuse of these rules may jeopardize the chances for endangered species recovery.

Trump and Executive Power

The Trump administration focused more intently on wholesale broad changes to the laws that underpin many of the regulations promulgated by agencies charged with environmental protection, thereby attempting to have more lasting influence over environmental regulations. The Endangered Species Act was no exception.

In 2019, the Trump administration promulgated several rules revising the ESA in an effort to better align economic impacts with conservation goals. As Interior Secretary David Bernhardt argued in a prepared statement, "An effectively administered Act ensures more resources can go where they will do the most good: on-the-ground conservation."[61] The rules allow for economic considerations to be vetted in listing species, though final listing decisions are still required to be based solely on scientific data. They also define the term *foreseeable future* as that which "extends only so far into the future as [the USFWS and NMFS] can reasonably determine that the conditions potentially posing a danger of extinction in the foreseeable future are probable,"[62] which critics contend will make it more difficult to list species based on the impacts of climate change.[63] Finally, the rules address the issue of critical habitat by identifying situations or circumstances where the USFWS may not need to designate critical habitat and reinstating an earlier regulation that areas occupied by protected species are evaluated for their habitat value before unoccupied areas.[64]

The new regulations were couched under President Trump's January 30, 2017, executive order to reduce regulation and control regulatory costs. Though Republicans largely supported the revisions, critics maintained that the regulations undermine the intent of the law. Congressman Raul Grijalva (D-AZ) attacked the actions of the administration, stating, "These rollbacks of the ESA are for one purpose only: more handouts to special interests that don't want to play by the rules and only want to line their pockets. This action by the Trump administration adds to their ongoing efforts to clear the way for oil and gas development without any regard for the destruction of wildlife and their habitats."[65]

Courts and the Endangered Species Act: Gray Wolf Recovery

The courts have been a significant actor in ESA challenges. The ESA includes a provision conferring on "any person" the capacity and right to enforce the act's mandates. Its purpose "is to provide for conservation, protection and propagation of endangered species of fish and wildlife by Federal action."[66] Various interpretations of environmental standing have held sway in the courts that either benefit environmental advocates or prevent their involvement in court action.[67] Given the vagueness in the original text of the ESA, the courts have been the primary venue in which the law has been clarified.[68] Nongovernmental organizations, such as the Center for Biological Diversity, Defenders of Wildlife, and others, have been instrumental in acting on behalf of endangered species through the courts. Famously, in *Bennett v. Spear*, questions of whether the citizen suit provision of the ESA grants standing to plaintiffs suing on the basis of economic harm resulting from alleged over-enforcement of the ESA were posed.[69] Although the court found that the petitioners had standing, it held that plaintiffs may not use the citizen suit provision for any purpose other than to compel nondiscretionary action. The overall effect of *Bennett* was to establish the ESA as a tool of both environmentalists and competing users of natural resources to challenge government action.

The policy history of gray wolf recovery in the Northern Rockies illustrates well the battles in the courts over the ESA. Wolves were added to the endangered species list in 1974. The ESA mandates that the USFWS prepare a recovery plan for any endangered species, and so with the addition of gray wolves to the list, a federal endeavor began to restore the species to the lower forty-eight states. The first step in this process was the appointment of a Northern Rocky Mountain Wolf Recovery Team in 1974, which was tasked with developing the recovery plan. After several years of revisions, in 1987 the USFWS adopted a final recovery plan for this region.[70] Two primary means of population recovery were identified: natural recolonization, where wolves would reenter recovery areas and establish themselves without human assistance, and reintroduction, where wolves would be relocated from source populations and released in the appropriate location identified by the recovery team. Initially, reintroduction was considered for only Yellowstone National Park; the central Idaho recovery area as well as western Montana were thought to be appropriate for natural recolonization, though wolves ultimately were reintroduced to central Idaho in 1996. The recovery plan also suggested the creation of a control plan to address wolf-livestock conflicts as well as ongoing research and monitoring over the recovery period. Finally, the plan established biological recovery goals for the species. Specifically, the northern Rocky Mountain wolf would be removed from the endangered species list once ten breeding pairs of wolves and one hundred individuals had been documented in each of the recovery zones for at least three successive years. Once the population reached this recovery target, the delisting process would be initiated.[71]

Under the National Environmental Policy Act, the USFWS was directed by Congress to prepare an environmental impact statement for the reintroduction alternative proposed in the recovery plan. It took another eight years for the USFWS to issue the final environmental impact statement, which identified reintroduction as the preferred alternative. One of the key compromises in the final environmental impact statement was the invocation of the ESA's 10(j) rule, discussed earlier, which allows the USFWS to categorize a reintroduced species as an "experimental and non-essential population," to provide for greater management flexibility. In practice, it provided managers with recourse to address conflicts on public or private lands; for example, if individual or packs of wolves harassed or killed livestock, they could be lethally managed.[72]

In January 1995, federal officials released fourteen wolves into Yellowstone National Park and fifteen to central Idaho.[73] In 2002, data collected through the monitoring of radio-collared wolves indicated the biological goals of the recovery plan had been met, and various stakeholders—particularly in the agricultural and outfitter sectors—called for the delisting process to begin. The first wolf delisting rule was published by the USFWS in March 2008, after the agency approved management plans for the three states in the recovery zone—Montana, Idaho, and Wyoming—that would ensure the long-term maintenance of wolf populations in accordance with the original recovery goals.

The decision of a subsequent lawsuit by environmental groups reinstated protection in July 2008 in all three states due to the inadequacy of Wyoming's wolf management plan.[74] The USFWS responded by delisting wolves in only Idaho and Montana in March 2009. Litigated yet again by environmental groups in June 2009, this partial delisting decision was challenged on the grounds that wolf populations could not be delisted in some states and not others.[75] Wolves returned yet again to the endangered species list in all three states in August 2010. The political and legal tug-of-war tapered somewhat in early 2011, when the USFWS delisted the wolf populations in Idaho and Montana as well as in other states that had wolves present due to dispersal but were not considered part of the recovery zone: Oregon, Washington, and Utah.[76] Though this move was technically a partial delisting decision, Congress had directed the USFWS to proceed with delisting through a rider in an appropriations bill, discussed earlier. For the first time in the history of the Endangered Species Act, delisting was decided by Congress, which set a precedent for delisting of endangered species by political action.[77] Several environmental groups litigated this action unsuccessfully in *Alliance for the Wild Rockies v. Salazar*, which "paved the way for further Congressional encroachment on the judiciary."[78]

The USFWS continued to work with Wyoming and approved the state's wolf management plan in October 2011. On March 8, 2012, Governor Matt Mead signed Wyoming's wolf management bill into law, and the USFWS turned management of Wyoming's wolf packs over to the state on September 30, 2012. Dan Ashe, the director of the USFWS at the time, said in a June 2013 blog post announcing the final delisting rule, "We have brought back this great icon of the American wilderness. And as we face today's seemingly insurmountable challenges, today's critical voices, today's political minefields, let this success be a reminder of what we can accomplish. We can work conservation miracles, because we have. The gray wolf is proof."[79]

Nonetheless, environmental groups contested the decision to remove Wyoming's wolves from the endangered species list after a ninety-day waiting period. In September 2014, a U.S. District Court decision in Washington, DC, returned Wyoming's wolves to the endangered species list and the USFWS resumed management once again.[80] This decision was primarily based on Wyoming's lack of written commitment in its management plan to maintain a population buffer above the recovery goals. This decision essentially reinforced USFWS's justification of its decision to remove wolves from the endangered species list, yet required Wyoming to maintain a specific population buffer that would ensure the state would not let its wolf numbers fall below the recovery goals and thus trigger a return to the endangered species list. From September 2014 through March 2017, wolf management in Wyoming remained under the jurisdiction of the USFWS. The 2014 ruling was appealed, however, and in March 2017, the U.S. Court of Appeals for the DC Circuit reversed the district court's decision and removed ESA protections from wolves in Wyoming. The USFWS subsequently passed the delisting rule for Wyoming's wolves in April 2017.

The court cases described above illustrate the role of litigation in clarifying important components of the ESA and balancing competing interests at stake. The cases have importance beyond just the case of the gray wolf; the USFWS solicited public comment based on the reasoning contained in one court case decided in 2017 regarding wolf delisting for a grizzly bear delisting decision.[81] Furthermore, the delisting of wolves in Montana and Idaho via congressional budget rider illustrated the potential for future congressional intervention in judicial decisions regarding the ESA.

States, Stakeholders, and the ESA: Sage-Grouse Conservation

In contrast to the ongoing litigation surrounding wolf management and policy, the example of greater sage-grouse conservation in the western United States illustrates an effort to deliberately and collaborative preclude the species from ESA listing by engaging states as well as conservation nongovernmental organizations and private citizens to develop sage-grouse conservation plans. Sage-grouse conservation efforts were novel in that the federal agencies directly engaged with the states to address management.

In 2010, the USFWS issued an ESA listing determination of "warranted but precluded" for the greater sage-grouse (*Centrocercus urophasianus*), meaning that the birds' population trends indicated they should be listed, but given resource constraints, other species had priority. A year later, in December 2011, the Obama administration provided eleven western states with sage-grouse populations the opportunity to develop state-specific management plans. If the USFWS then approved these plans, states would be exempted from the Bureau of Land Management's national sage-grouse management strategy for federally managed lands.[82] States were motivated to partner with the USFWS in developing individualized plans given the potential economic impacts to the state should the sage-grouse be listed as an endangered species. For example, Wyoming actively engaged in the sage-grouse planning process because of the overlap of sage-grouse habitat with areas of concentrated resource extraction.[83]

The collaborative processes between the USFWS and the states were individualized to suit the context of the problem locally or regionally. California and Nevada worked together to develop their conservation plans, because the population of sage-grouse in this region spans state lines. Three committees oversaw different aspects of the process under a jointly developed "partnership for collaboration," which "empowered local stakeholders, fostered coordination among ownership boundaries, and ensured the best available science was used in decision making." This effort engaged state wildlife and federal land management agencies as well as biologists, private landowners, citizens, tribes, the Department of Defense (which has land management oversight on some federal lands throughout Nevada), and nongovernmental organizations.[84] Part of the outcome of the sage-grouse planning process was the identification of about 10.7 million acres of high value sage-grouse habitat, termed "Sagebrush Focal Areas," where oil and gas drilling was limited.

In 2015, then secretary of the interior Sally Jewell lauded greater sage-grouse recovery planning in the western United States as a "truly a historic effort—one that represents extraordinary collaboration across the American West. . . . It demonstrates that the ESA is an effective and flexible tool and a critical catalyst for conservation—ensuring that future generations can enjoy the diversity of wildlife that we do today. The epic conservation effort will benefit westerners and hundreds of species that call this iconic landscape home, while giving states, businesses and communities the certainty they need to plan for sustainable economic development."[85] However, following the 2016 presidential election, President Trump adopted a platform of "energy dominance," which involved increasing oil and gas operations nationally, and particularly on public lands. The sagebrush focal areas identified in 2015 were reduced from 10.7 million acres to 1.8 million acres, which was a significant gain for oil and gas development interests.[86] However, this move was blocked by the district court in Idaho in October 2019; this decision reinstated the 2015 Obama-era protections for sage-grouse habitat.[87]

The example of sage-grouse collaboration in the western United States illustrates that despite efforts to find common-ground solutions that both protect species and address the needs of stakeholders, politics can still trump collaborative outcomes. In this case, the influence and priorities of the executive office sought to counter the conservation priorities of the agencies and stakeholders in the western states involved with collaborative efforts to conserve sage-grouse and mitigate the potential for a future endangered species listing.

As discussed earlier, one key role for states involves implementing regulations, which can help create laws through the on-the-ground actions and interpretations of government personnel. ESA delegates authority for implementation to the USFWS and the NMFS. Efforts persist to delegate authority, or primacy, to state agencies to oversee the ESA; however, analyses suggest that state endangered species laws are much weaker than the ESA and therefore insufficient to address conservation and recovery goals.[88] Furthermore, allocating ESA responsibilities to the states would also necessitate a significantly higher financial outlay on the part of the state government than most currently allocate to endangered species recovery.[89]

The USFWS has its own law enforcement branch that serves the role of environmental inspector, as described above. This office's mission "is to protect wildlife and plant resources through the effective enforcement of federal laws, . . . combat wildlife trafficking, help recover endangered species, conserve migratory birds, preserve wildlife habitat, safeguard fisheries, prevent the introduction and spread of invasive species, and promote international wildlife conservation."[90] For example, if an endangered species is poached, USFWS wildlife managers would call law enforcement to conduct the investigation.

Beyond the direct arm of law enforcement, the USFWS also implements post-delisting monitoring to ensure that the populations of species delisted due to biological recovery do not decline such that relisting would become necessary. Section 4(g) of the ESA (added via amendments in 1988) requires the USFWS to work with the states where recovered species occur to monitor their populations for a minimum of five years after delisting. Following this five-year oversight period, the USFWS can decide whether to relist, continue to monitor, or step back entirely from management. The USFWS must consult with states in developing post-delisting monitoring plans; other stakeholders—including other programs of the USFWS, other federal agencies, tribes, and landowners—are strongly encouraged to be involved with this planning as well.[91]

Concluding Thoughts

As this chapter outlines, major institutional actors vie for influence over rulemaking outcomes. Once a rule is passed, presidents, Congress, and the judiciary all play roles in the jockeying that takes place. Presidents may attempt to influence rulemaking through their power of appointment and the use of executive orders requiring specific policy priorities to be considered in rule promulgation (e.g., cost-benefit analysis). Congress and presidents also wrestle over their legislative priorities to enact potential changes to rules through tools such as the Congressional Review Act, budget riders, and appropriations to the executive branch. While these are primarily the tools of influence for Congress, presidents can choose to support them and sign them into law or attempt to stymie these changes with veto threats. Finally, as a last resort, advocates both inside and outside formal institutions can seek remedies in the judiciary to overturn rules with which they disagree. This chapter also highlights the key roles states play in interpreting, monitoring, and implementing regulations—all roles that can modify a final rule. This back-and-forth can lead to years of conflict and appeals for changes from the advocates on competing sides of policy issues. The tools described in this chapter provide some insight into the seemingly never-ending rulemaking process that is far from final after a rule is finalized. Most importantly, this chapter recognizes the importance of state-level actors, which is the focus of Chapters 5, 6, and 7.

Notes

1. Lawrence Hurley, "Dusky Gopher Frog Suffers Setback in U.S. Supreme Court Ruling," *Reuters*, November 27, 2018, https://www.reuters.com/article/us-usa-court-frog/dusky-gopher-frog-suffers-setback-in-u-s-supreme-court-ruling-idUSKCN1NW1P3.

2. U.S. Fish and Wildlife Service, Interior; National Marine Fisheries Service, National Oceanic and Atmospheric Administration, Commerce, Proclamation, "Endangered and Threatened Wildlife and Plants; Regulations for Interagency Cooperation," *Federal Register* 84, no. 166 (August 27, 2019): 44976–45016, https://www.govinfo.gov/content/pkg/FR-2019-08-27/pdf/2019-17517.pdf.

3. Deserai A. Crow, Elizabeth A. Albright, and Elizabeth A. Koebele, "Environmental Rulemaking across States: Process, Procedural Access, and Regulatory

Influence," *Environment and Planning C: Government and Policy* 34, no. 7 (2016): 1222–1240; Deserai A. Crow, Elizabeth A. Albright, and Elizabeth A. Koebele, "Public Information and Regulatory Processes: What the Public Knows and Regulators Decide," *Review of Policy Research* 33, no. 1 (2016): 90–109; Christine A. Kelleher and Susan Webb Yackee, "Who's Whispering in Your Ear? The Influence of Third Parties over State Agency Decisions," *Political Research Quarterly* 59, no. 4 (2006): 629–643; William F. West and Connor Raso, "Who Shapes the Rulemaking Agenda? Implications for Bureaucratic Responsiveness and Bureaucratic Control," *Journal of Public Administration Research and Theory* 23, no. 3 (2013): 495–519; Jason Webb Yackee and Susan Webb Yackee, "A Bias towards Business? Assessing Interest Group Influence on the U.S. Bureaucracy," *Journal of Politics* 68 (2006): 128–139; Susan Webb Yackee, "Sweet-Talking the Fourth Branch: The Influence of Interest Group Comments on Federal Agency Rulemaking," *Journal of Public Administration Research and Theory* 16 (2006): 103–124.

4. David E. Lewis, *The Politics of Presidential Appointments: Political Control and Bureaucratic Performance* (Princeton, NJ: Princeton University Press, 2010).

5. Jerry W. Calvert, "Party Politics and Environmental Policy," *Environmental Politics and Policy* (1989): 158–178; Riley E. Dunlap and Rik Scarce, "Poll Trends: Environmental Problems and Protection," *Public Opinion Quarterly* 55, no. 4 (1991): 651–672; Riley E. Dunlap, "Polls, Pollution, and Politics Revisited Public Opinion on the Environment in the Reagan Era," *Environment: Science and Policy for Sustainable Development* 29, no. 6 (1987): 6–37.

6. Terry M. Moe, "Regulatory Performance and Presidential Administration," *American Journal of Political Science* (1982): 197–224; David E. Lewis, "The Contemporary Presidency: The Personnel Process in the Modern Presidency," *Presidential Studies Quarterly* 42, no. 3 (2012): 577–596.

7. Brad Plumer and Coral Davenport, "Science under Attack: How Trump Is Sidelining Researchers and Their Work," *New York Times*, December 28, 2019, https://nytimes.com/2019/12/28/climate/trump-administration-war-on-science.html.

8. Steven M. Johnson, "Disclosing the President's Role in Rulemaking: A Critique of the Reform Proposals," *Catholic University Law Review* 60 (2010): 1003.

9. Lewis, *The Contemporary Presidency*.

10. Morton Rosenberg, "Presidential Control of Agency Rulemaking: An Analysis of Constitutional Issues That May Be Raised by Executive Order 12,291," *Arizona Law Review* 23 (1981): 1199.

11. Peter M. Shane, "Presidential Regulatory Oversight and the Separation of Powers: The Constitutionality of Executive Order No. 12,291," *Arizona Law Review* 23 (1981): 1235.

12. Erik D. Olson, "The Quiet Shift of Power: Office of Management and Budget Supervision of Environmental Protection Agency Rulemaking under Executive Order 12,291," *Virginia Journal of Natural Resources Law* 4 (1984): 1.

13. Robert W. Hahn and Cass R. Sunstein, "A New Executive Order for Improving Federal Regulation? Deeper and Wider Cost-Benefit Analysis," *University of Pennsylvania Law Review* 150, no. 5 (2002): 1489–1552.

14. Lisa Friedman, "Trump Weakens Major Conservation Law to Speed Construction Permits," *New York Times*, July 15, 2020, https://www.nytimes.com/2020/07/15/climate/trump-environment-nepa.html.

15. Philip J. Harter, "Executive Oversight of Rulemaking: The President Is No Stranger. A Symposium on Administrative Law—The Uneasy Constitutional Status of the Administrative Agencies: Part II—Presidential Oversight of Regulatory Decisionmaking: Commentary," *American University Law Review* 2 (1986): 557–572.

16. Harter, "Executive Oversight"; Kenneth Lowande, "Politicization and Responsiveness in Executive Agencies," *Journal of Politics* 81, no. 1 (2019): 33–48.

17. Trevor R. S. Allan, "Legislative Supremacy and the Rule of Law: Democracy and Constitutionalism," *Cambridge Law Journal* 44 (1985): 111; Calvin R. Massey, "The Locus of Sovereignty: Judicial Review, Legislative Supremacy, and Federalism in the Constitutional Traditions of Canada and the United States," *Duke Law Journal* 1990, no. 6 (1990): 1229–1310; Charles Howard McIlwain, *The High Court of Parliament and Its Supremacy: An Historical Essay on the Boundaries between Legislation and Adjudication in England*, vol. 105 (New Haven, CT: Yale University Press, 1910).

18. Abner J. Mikva, "Congress: The Purse, the Purpose, and the Power," *Georgia Law Review* 21 (1986): 1; Joachim Wehner, "Assessing the Power of the Purse: An Index of Legislative Budget Institutions," *Political Studies* 54, no. 4 (2006): 767–785.

19. Lance T. LeLoup, *The Fiscal Congress: Legislative Control of the Budget* (Westport, CT: Greenwood, 1980).

20. Jeb Barnes, *Overruled? Legislative Overrides, Pluralism, and Contemporary Court-Congress Relations* (Palo Alto, CA: Stanford University Press, 2004).

21. Morton Rosenberg, "Whatever Happened to Congressional Review of Agency Rulemaking: A Brief Overview, Assessment, and Proposal for Reform," *Administrative Law Review* 4 (1999): 1051–1092.

22. Morton Rosenberg, "Congressional Review of Agency Rulemaking: An Update and Assessment of the Congressional Review Act after a Decade," *Congressional Research Service, Library of Congress*, 2008.

23. James T. O'Reilly, "FDA Rulemaking after the 104th Congress: Major Rules Enter the Twilight Zone of Review," *Food and Drug Law Journal* 4 (1996): 677–696.

24. Julie A. Parks, "Lessons in Politics: Initial Use of the Congressional Review Act Comment," *Administrative Law Review* 1 (2003): 187–210.

25. Daniel Cohen and Peter L. Strauss, "Congressional Reviews of Agency Regulations Recent Developments—Regulatory Reform and the 104th Congress," *Administrative Law Review* 1 (1997): 95–110.

26. Thomas T. Holyoke, Heath Brown, and Jeffrey R Henig, "Shopping in the Political Arena: Strategic State and Local Venue Selection by Advocates," *State and Local Government Review* 44, no. 1 (2012): 9–20.

27. Frank R. Baumgartner and Bryan D. Jones, *Agendas and Instability in American Politics* (Chicago: University of Chicago Press, 1993).

28. Holyoke et al., "Shopping in the Political Arena."

29. Thomas T. Holyoke, "Choosing Battlegrounds: Interest Group Lobbying across Multiple Venues," *Political Research Quarterly* 56, no. 3 (2003): 325–336.

30. Maxwell L. Stearns, "From Lujan to Laidlaw: A Preliminary Model of Environmental Standing Citizen Suits and the Future of Standing in the 21st Century: From Lujan to Laidlaw and Beyond," *Duke Environmental Law and Policy Forum* 2 (2000): 321–388.

31. Ibid.

32. Jonathan H. Adler, "Stand or Deliver: Citizen Suits, Standing, and Environmental Protection," *Duke Environmental Law and Policy Forum* 12 (2001): 39. Lujan v. Defenders of Wildlife, 504 U.S. 555 (1992); Friends of the Earth, Inc. v. Laidlaw Environmental Services, Inc., 528 U.S. 167 (2000).

33. Jonathan H. Adler, "When Is Two a Crowd—The Impact of Federal Action on State Environmental Regulation," *Harvard Environmental Law Review* 31 (2007): 67.

34. Norman Vig and Michael E. Kraft, *Environmental Policy: New Directions for the Twenty-First Century* (Washington, DC: CQ Press, 2018).

35. Crow et al., "Environmental Rulemaking"; Crow et al., "Public Information"; Michael Steeves, "The EPA's Proposed CAFO Regulations Fall Short of Ensuring the Integrity of Our Nation's Waters," *Journal of Land Resources and Environmental Law* 22 (2002): 367–397.

36. Elizabeth A. Koebele, "Cross-Coalition Coordination in Collaborative Environmental Governance Processes," *Policy Studies Journal* (2019): doi:10.1111/psj.12306; Elizabeth A. Koebele, "Assessing Outputs, Outcomes, and Barriers in Collaborative Water Governance: A Case Study," *Journal of Contemporary Water Research and Education* 155, no. 1 (2015): 63–72; Thomas M. Koontz and Elizabeth Moore Johnson, "One Size Does Not Fit All: Matching Breadth of Stakeholder Participation to Watershed Group Accomplishments," *Policy Sciences* 37, no. 2 (2004): 185–204; Katrina Smith Korfmacher and Thomas M. Koontz, "Collaboration, Information, and Preservation: The Role of Expertise in Farmland and Preservation Task Forces," *Policy Sciences* 36, no. 3–4 (2003): 213–236; Mark Lubell, "Collaborative Environmental Institutions: All Talk and No Action?" *Journal of Policy Analysis and Management* 23, no. 3 (2004): 549–573; Elizabeth A. Moore and Tomas M Koontz, "Research Note a Typology of Collaborative Watershed Groups: Citizen-Based, Agency-Based, and Mixed Partnerships," *Society and Natural Resources* 16, no. 5 (2003): 451–460; Paul A. Sabatier, Will Focht, Mark Lubell, Zev Trachtenberg, Arnold Vedlitz, Marty Matlock, Michael E Kraft, and Sheldon Kamieniecki, *Swimming Upstream: Collaborative Approaches to Watershed Management* (Cambridge, MA: MIT Press, 2005).

37. David Cameron and Richard Simeon, "Intergovernmental Relations in Canada: The Emergence of Collaborative Federalism," *Publius: The Journal of Federalism* 32, no. 2 (2002): 49–72; Naim Kapucu and Vener Garayev, "Designing, Managing, and Sustaining Functionally Collaborative Emergency Management Networks," *American Review of Public Administration* 43, no. 3 (2012): 312–330; Mark Lubell, "Collaborative Institutions, Belief-Systems, and Perceived Policy Effectiveness," *Political Research Quarterly* 56, no. 3 (2003): 309–323; Nicola Ulibarri, "Tracing Process to Performance of Collaborative Governance: A Comparative Case Study of Federal Hydropower Licensing," *Policy Studies Journal* 43, no. 2 (2015): 283–308.

38. Peter L. Hupe and Michael J. Hill, "'And the Rest Is Implementation.' Comparing Approaches to What Happens in Policy Processes beyond *Great Expectations*," *Public Policy and Administration* 31, no. 2 (2016): 103–121; Peter J. May and Soren C. Winter, "Politicians, Managers, and Street-Level Bureaucrats: Influences on Policy Implementation," *Journal of Public Administration Research and Theory* 19, no. 3 (2009): 453–476; Jeffrey L. Pressman and Aaron B. Wildavsky, *Implementation: How Great Expectations in Washington Are Dashed in Oakland: Or, Why It's Amazing That Federal Programs Work at All, This Being a Saga of the Economic Development Administration as Told by Two Sympathetic Observers Who Seek to Build Morals on a Foundation of Ruined Hopes* (Berkeley: University of California Press, 1973).

39. May and Winter, "Politicians, Managers, and Street-Level Bureaucrats."

40. John P. Dwyer, "The Role of State Law in an Era of Federal Preemption: Lessons from Environmental Regulation," *Law and Contemporary Problems* 60 (1997): 203.

41. B. Dan Wood, "Modeling Federal Implementation as a System: The Clean Air Case," *American Journal of Political Science* 36, no. 1 (1992): 40–67.

42. Michelle C. Pautz, "Trust between Regulators and the Regulated: A Case Study of Environmental Inspectors and Facility Personnel in Virginia," *Politics and Policy* 37, no. 5 (2009): 1047–1072.

43. Michelle C. Pautz, "Next-Generation Environmental Policy and the Implications for Environmental Inspectors: Are Fears of Regulatory Capture Warranted?" *Environmental Practice* 12, no. 3 (2010); Michelle C. Pautz and Marcy H. Schnitzer, "Policymaking from Below: The Role of Environmental Inspectors and Publics," *Administrative Theory and Praxis* 30, no. 4 (2008): 450–475.

44. Stephen Fineman, "Street-Level Bureaucrats and the Social Construction of Environmental Control," *Organization Studies* 19, no. 6 (1998): 953–974.

45. Wayne B. Gray and Jay P. Shimshack, "The Effectiveness of Environmental Monitoring and Enforcement: A Review of the Empirical Evidence," *Review of Environmental Economics and Policy* 5, no. 1 (2011): 3–24.

46. According to the ESA, an "endangered" species is any one "in danger of extinction throughout all or a significant portion of its range," while a "threatened" species is "any species which is likely to become an endangered species within the foreseeable future throughout all or a significant portion of its range."

For the purposes of this chapter, we refer to both of these categories with the term *protected species*.

47. Daniel J. Rohlf, "Section 4 of the Endangered Species Act: Top Ten Issues for the Next Thirty Years," *Environmental Law* 34, no. 2 (2004): 483.

48. Jacob W. Malcom and Ya-Wei Li, "Data Contradict Common Perceptions about a Controversial Provision of the US Endangered Species Act," *Proceedings of the National Academy of Sciences* 112, no. 52 (2015): 15844–15849.

49. Endangered and Threatened Wildlife and Plants; Regulations for Interagency Cooperation. 50 CFR Part 402. *Federal Register* 84, no. 166 (2019): 44976–45016.

50. P. A. Sheikh, E. H. Ward, and R. E. Crafton, *Final Rules Changing Endangered Species Act Regulations*, ed. Congressional Research Service. (Washington, DC, 2019).

51. Ward Sinclair, "'Pork Panic' Sweeping Congress in Wake of Darter's Rescue," *Washington Post*, June 28, 1978.

52. David R. Mayhew, *Congress: The Electoral Connection* (New Haven, CT: Yale University Press, 1974).

53. M. Lynne Corn and Alexandra M. Wyatt, "The Endangered Species Act: A Primer," *Congressional Research Service*, September 8, 2016, https://fas.org/sgp/crs/misc/RL31654.pdf.

54. Michael J. Bean and Melanie J. Rowland, *The Evolution of National Wildlife Law* (Westport, CT: Greenwood, 1997), 232–233.

55. Joseph A. Davis, "Environment 1982: Overview," *CQ Almanac* 1982, no. 38 (1983): 423–424.

56. "No Consensus Reached on Forest Protection," *CQ Almanac* 1992, no. 48 (1993): 277–280.

57. Judith A. Layzer, *Open for Business: Conservatives' Opposition to Environmental Regulation* (Cambridge, MA: MIT Press, 2012).

58. Mary Clare Jalonick, "Environmental Panels' Chairmen Chip Away at Endangered Species Act, Refocusing Resources and Definitions," *CQ Weekly* (March 27, 2004): 756.

59. Layzer, *Open for Business*.

60. Phil Taylor and Jean Chemnick, "House Panel Advances Rider-Laden Interior, EPA Bill," *Greenwire*, July 7, 2011.

61. U.S. Department of the Interior, "Trump Administration Improves the Implementing Regulations of the Endangered Species Act," August 12, 2019, https://www.doi.gov/pressreleases/endangered-species-act.

62. "Endangered and Threatened Wildlife and Plants; Regulations for Listing Species and Designating Critical Habitat," 50 CFR Part 424, *Federal Register* 84, no. 166 (2019). 45020–45053.

63. Lisa Friedman, "U.S. Significantly Weakens Endangered Species Act," *New York Times*, August 12, 2019, https://www.nytimes.com/2019/08/12/climate/endangered-species-act-changes.html?searchResultPosition=7.

64. 84 FR 45020; Sheikh et al., *Final Rules*.

65. Office of Congressman Raul M. Grijalva, "Chair Grijalva: President Trump' s Assault on Endangered Species Will Add to Extinction Crisis as a Favor to Industry," 2019, https://grijalva.house.gov/press-releases/chair-grijalva -president-trumps-assault-on-endangered-species-will-add-to-extinction-crisis -as-a-favor-to-industry/.

66. Raymond A. Just, "Intergenerational Standing under the Endangered Species Act: Giving Back the Right to Biodiversity after Lujan v. Defenders of Wildlife," *Tulane Law Review* 71 (1996): 597.

67. Philip Weinberg, "Are Standing Requirements Becoming a Great Barrier Reef against Environmental Actions?" *NYU Environmental Law Journal* 7 (1999): 1.

68. Vig and Kraft, *Environmental Policy*.

69. Bennett v. Spear 520 U.S. 154, 176 (1997).

70. The USFWS also identified a second distinct population segment of wolves, the Western Great Lakes, and a recovery plan was adopted in 1978 and revised in 1992. This plan required a "stable or growing" Minnesota population (of around fifteen hundred animals), and essentially a Wisconsin-Michigan population of one hundred animals.

71. U.S. Fish and Wildlife Service, ed., *Northern Rocky Mountain Wolf Recovery Plan* (Denver, CO, 1987).

72. Endangered and Threatened Wildlife and Plants; Final Rule Designating the Northern Rocky Mountain Population of Gray Wolf as a Distinct Population Segment and Removing This Distinct Population Segment from the Federal List of Endangered and Threatened Wildlife, 50 CFR Part 17, *Federal Register* 73, no. 39 (2008), 10514–10560.

73. Edward E. Bangs and Steven H. Fritts, "Reintroducing the Gray Wolf to Central Idaho and Yellowstone National Park," *Wildlife Society Bulletin* 24, no. 3 (1996): 402–413; Douglas W. Smith, Wayne G. Brewster, and Edward E. Bangs, "Wolves in the Greater Yellowstone Ecosystem: Restoration of a Top Carnivore in a Complex Management Environment," in *Carnivores in Ecosystems: The Yellowstone Experience*, ed. Tim W. Clark, A. Peyton Curlee, Steven C. Minta, and Peter M. Karieva, 103–126 (New Haven, CT: Yale University Press, 1999).

74. *Defenders of Wildlife et al. v. Hall et al.* 2008. 565 F. Supp. 2d 1160. U.S. District Court, Missoula, Montana.

75. *Defenders of Wildlife v. Salazar.* 2010. 729 F. Supp. 2d 1207. U.S. District Court, Missoula, Montana.

76. Endangered and Threatened Wildlife and Plants; Reissuance of Final Rule to Identify the Northern Rocky Mountain Population of Gray Wolf as a Distinct Population Segment and to Revise the List of Endangered and Threatened Wildlife, 50 CFR Part 17, *Federal Register* 76, no. 87 (2011), 25590–25592.

77. Felicity Barringer and John M. Broder, "Congress, in a First, Removes an Animal from the Endangered Species List," *New York Times*, April 12, 2011; Emily E. Cathcart, "All Bark and No Bite: Nonspecific Magic Words Sweep Aside Constitutional Concerns and Remove the Northern Rocky Mountain Gray Wolf from Endangered Species Act Protection in Alliance for the Wild

Rockies v. Salazar," *Villanova Environmental Law Journal* 24 (2013): 253–285, doi:10.1525/sp.2007.54.1.23.

78. Cathcart, "All Bark and No Bite," 253; Alliance for Wild Rockies v. Salazar (Wild Rockies II) 672 F.3d 1170 (9th Cir. 2012).

79. Dan Ashe, "Director's Corner: Gray Wolves Are Recovered; Next Up, the Mexican Wolf," U.S. Fish and Wildlife Service, August 12, 2019, https://www.fws.gov/director/dan-ashe/index.cfm/2013/6/7/Gray-wolves-are-recovered-next-up-the-Mexican-wolf.

80. *Defenders of Wildlife, et al., v. Jewell, et al.* 2014. 68 F. Supp. 3d 193. U.S. District Court for the District of Columbia.

81. *Humane Society of the U.S. vs. Zinke.* 2017. U.S. Court of Appeals for the District of Columbia. No. 15-5041.

82. Cally Younger and Sam Eaton, "Lessons Learned from the Greater Sage-Grouse Land Use Planning Effort," *Idaho Law Review* 53, no. 2 (2018): 373.

83. Peter D. Stahl and Michael F. Curran, "Collaborative Efforts Towards Ecological Habitat Restoration of a Threatened Species, Greater Sage-Grouse, in Wyoming, USA," *Land Reclamation in Ecological Fragile Areas—Proceedings of the 2nd International Symposium on Land Reclamation and Ecological Restoration*, LRER 2017, 251–254, https://doi.org/10.1201/9781315166582-53.

84. Alison L. Duvall, Alexander L. Metcalf, and Peter S. Coates, "Conserving the Greater Sage-Grouse: A Social-Ecological Systems Case Study from the California-Nevada Region," *Rangeland Ecology and Management* 70, no. 1 (2017): 134, doi:10.1016/j.rama.2016.08.001.

85. U.S. Department of the Interior, "Historic Conservation Campaign Protects Greater Sage-Grouse." U.S. Fish and Wildlife Service, September 22, 2015, https://www.doi.gov/pressreleases/historic-conservation-campaign-protects-greater-sage-grouse.

86. Coral Davenport, "Trump Administration Loosens Sage Grouse Protections, Benefiting Oil Companies," *New York Times*, March 15, 2019.

87. Lisa Friedman, "Court Blocks Trump's Plan to Ease Bird Protections on Oil Lands," *New York Times*, October 16, 2019.

88. Two states, Wyoming and West Virginia, do not have state endangered species laws.

89. Alejandro E. Camacho, Michael Robinson-Dorn, Asena Cansu Yildiz, and Tara Teegarden, "Assessing State Laws and Resources for Endangered Species Protection," *Environmental Law Reporter* 47, no. 10 (2017): 10837–10844.

90. U.S. Fish and Wildlife Service, "Mission," 2020, https://www.fws.gov/le/.

91. U.S. Fish and Wildlife Service and National Marine Fisheries Service. Post-Delisting Monitoring Plan Guidance under the Endangered Species Act. 2018.

Suggested Readings

Barnes, Jeb. *Overruled? Legislative Overrides, Pluralism, and Contemporary Court-Congress Relations*. Palo Alto, CA: Stanford University Press, 2004.

Doub, Peyton J. *The Endangered Species Act: History, Implementation, Successes, and Controversies*. Boca Raton, FL: CRC Press, 2012.

Fiorino, Daniel J. *The New Environmental Regulation*. Cambridge, MA: MIT Press, 2006.

Johnson, Steven M. "Disclosing the President's Role in Rulemaking: A Critique of the Reform Proposals." *Catholic University Law Review* 60 (2010): 1003–1044.

Parks, Julie A. "Lessons in Politics: Initial Use of the Congressional Review Act Comment." *Administrative Law Review* 1 (2003): 187–210.

Pautz, Michelle C. "Next-Generation Environmental Policy and the Implications for Environmental Inspectors: Are Fears of Regulatory Capture Warranted?" *Environmental Practice* 12, no. 3 (2010): 247–259.

Bibliography

Adler, Jonathan H. "Stand or Deliver: Citizen Suits, Standing, and Environmental Protection." *Duke Environmental Law and Policy Forum* 12 (2001): 39–84.

———. "When Is Two a Crowd—The Impact of Federal Action on State Environmental Regulation." *Harvard Environmental Law Review* 31 (2007): 67–114.

Allan, Trevor R. S. "Legislative Supremacy and the Rule of Law: Democracy and Constitutionalism." *Cambridge Law Journal* 44 (1985): 111–143.

Ashe, Dan. "Director's Corner: Gray Wolves Are Recovered; Next up, the Mexican Wolf." U.S. Fish and Wildlife Service, August 12, 2019. Available at https://www.fws.gov/director/dan-ashe/index.cfm/2013/6/7/Gray-wolves-are-recovered-next-up-the-Mexican-wolf.

Bangs, Edward E., and Steven H. Fritts. "Reintroducing the Gray Wolf to Central Idaho and Yellowstone National Park." *Wildlife Society Bulletin* 24, no. 3 (1996): 402–413.

Barnes, Jeb. *Overruled? Legislative Overrides, Pluralism, and Contemporary Court-Congress Relations*. Palo Alto, CA: Stanford University Press, 2004.

Barringer, Felicity, and John M. Broder. "Congress, in a First, Removes an Animal from the Endangered Species List." *New York Times*, April 12, 2011.

Baumgartner, Frank R., and Bryan D. Jones. *Agendas and Instability in American Politics*. Chicago: University of Chicago Press, 1993.

Bean, Michael J., and Melanie J. Rowland. *The Evolution of National Wildlife Law*. Westport, CT: Greenwood, 1997.

Calvert, Jerry W. "Party Politics and Environmental Policy." *Environmental Politics and Policy* (1989): 158–178.

Camacho, Alejandro E., Michael Robinson-Dorn, Asena Cansu Yildiz, and Tara Teegarden. "Assessing State Laws and Resources for Endangered Species Protection." *Environmental Law Reporter* 47, no. 10 (2017): 10837–10844.

Cameron, David, and Richard Simeon. "Intergovernmental Relations in Canada: The Emergence of Collaborative Federalism." *Publius: The Journal of Federalism* 32, no. 2 (2002): 49–72.

Cathcart, Emily E. "All Bark and No Bite: Nonspecific Magic Words Sweep Aside Constitutional Concerns and Remove the Northern Rocky Mountain Gray Wolf from Endangered Species Act Protection in Alliance for the Wild Rockies v. Salazar." *Villanova Environmental Law Journal* 24 (2013): 253–285.

Cohen, Daniel, and Peter L. Strauss. "Congressional Reviews of Agency Regulations Recent Developments—Regulatory Reform and the 104th Congress." *Administrative Law Review* 1 (1997): 95–110.

Corn, M. Lynne, and Alexandra M. Wyatt. "The Endangered Species Act: A Primer." *Congressional Research Service*, September 8, 2016. Available at https://fas.org/sgp/crs/misc/RL31654.pdf.

Crow, Deserai A., Elizabeth A. Albright, and Elizabeth A. Koebele. "Environmental Rulemaking across States: Process, Procedural Access, and Regulatory Influence." *Environment and Planning C: Government and Policy* 34, no. 7 (2016): 1222–1240.

———. "Public Information and Regulatory Processes: What the Public Knows and Regulators Decide." *Review of Policy Research* 33, no. 1 (2016): 90–109.

Davenport, Coral. "Trump Administration Loosens Sage Grouse Protections, Benefiting Oil Companies." *New York Times*, March 15, 2019.

Davis, Joseph A. "Environment 1982: Overview." *CQ Almanac* 1982, no. 38 (1983): 423–424.

Defenders of Wildlife et al. v. Hall et al. 2008. 565 F. Supp. 2d 1160. U.S. District Court, Missoula, Montana.

Defenders of Wildlife et al. v. Jewell et al. 2014. 68 F. Supp. 3d 193. U.S. District Court for the District of Columbia.

Defenders of Wildlife v. Salazar. 2010. 729 F. Supp. 2d 1207. U.S. District Court, Missoula, Montana.

Dunlap, Riley E. "Polls, Pollution, and Politics Revisited Public Opinion on the Environment in the Reagan Era." *Environment: Science and Policy for Sustainable Development* 29, no. 6 (1987): 6–37.

Dunlap, Riley E., and Rik Scarce. "Poll Trends: Environmental Problems and Protection." *Public Opinion Quarterly* 55, no. 4 (1991): 651–672.

Duvall, Alison L., Alexander L. Metcalf, and Peter S. Coates. "Conserving the Greater Sage-Grouse: A Social-Ecological Systems Case Study from the California-Nevada Region." *Rangeland Ecology and Management* 70, no. 1 (2017): 129–140. Available at https://doi.org/10.1016/j.rama.2016.08.001.

Dwyer, John P. "The Role of State Law in an Era of Federal Preemption: Lessons from Environmental Regulation." *Law and Contemporary Problems* 60 (1997): 203–230.

Endangered and Threatened Wildlife and Plants; Final Rule Designating the Northern Rocky Mountain Population of Gray Wolf as a Distinct Population Segment and Removing This Distinct Population Segment from the Federal List of Endangered and Threatened Wildlife. 50 CFR Part 17. *Federal Register* 73, no. 39 (2008): 10514–10560.

Endangered and Threatened Wildlife and Plants; Regulations for Interagency Cooperation. 50 CFR Part 402. *Federal Register* 84, no. 166 (2019): 44976–45016.

Endangered and Threatened Wildlife and Plants; Reissuance of Final Rule to Identify the Northern Rocky Mountain Population of Gray Wolf as a Distinct Population Segment and to Revise the List of Endangered and Threatened Wildlife. 50 CFR Part 17. *Federal Register* 76, no. 87 (2011): 25590–25592.

Federal Register. "Endangered and Threatened Wildlife and Plants; Regulations for Listing Species and Designating Critical Habitat." 50 CFR Part 424. *Federal Register* 84, no. 166 (2019): 45020–45053.

Fineman, Stephen. "Street-Level Bureaucrats and the Social Construction of Environmental Control." *Organization Studies* 19, no. 6 (1998): 953–974.

Friedman, Lisa. "Court Blocks Trump's Plan to Ease Bird Protections on Oil Lands." *New York Times*, October 16, 2019.

———. "Trump Weakens Major Conservation Law to Speed Construction Permits," *New York Times*, July 15, 2020. Available at https://www.nytimes.com/2020/07/15/climate/trump-environment-nepa.html.

———. "U.S. Significantly Weakens Endangered Species Act." *New York Times*, August 12, 2019. Available at https://www.nytimes.com/2019/08/12/climate/endangered-species-act-changes.html?searchResultPosition=7.

Gray, Wayne B., and Jay P Shimshack. "The Effectiveness of Environmental Monitoring and Enforcement: A Review of the Empirical Evidence." *Review of Environmental Economics and Policy* 5, no. 1 (2011): 3–24.

Hahn, Robert W., and Cass R. Sunstein. "A New Executive Order for Improving Federal Regulation? Deeper and Wider Cost-Benefit Analysis." *University of Pennsylvania Law Review* 150, no. 5 (2002): 1489–1552.

Harter, Philip J. "Executive Oversight of Rulemaking: The President Is No Stranger. A Symposium on Administrative Law—The Uneasy Constitutional Status of the Administrative Agencies: Part II—Presidential Oversight of Regulatory Decisionmaking: Commentary." *American University Law Review* 2 (1986): 557–572.

Holyoke, Thomas T. "Choosing Battlegrounds: Interest Group Lobbying across Multiple Venues." *Political Research Quarterly* 56, no. 3 (2003): 325–336.

Holyoke, Thomas T., Heath Brown, and Jeffrey R Henig. "Shopping in the Political Arena: Strategic State and Local Venue Selection by Advocates." *State and Local Government Review* 44, no. 1 (2012): 9–20.

Humane Society of the U.S. vs. Zinke. 2017. U.S. Court of Appeals for the District of Columbia. No. 15-5041.

Hupe, Peter L., and Michael J. Hill. "'And the Rest Is Implementation.' Comparing Approaches to What Happens in Policy Processes beyond *Great Expectations*." *Public Policy and Administration* 31, no. 2 (2016): 103–121.

Hurley, Lawrence. "Dusky Gopher Frog Suffers Setback in U.S. Supreme Court Ruling." *Reuters*, November 27, 2018. Available at https://www.reuters

.com/article/us-usa-court-frog/dusky-gopher-frog-suffers-setback-in-u-s-supreme-court-ruling-idUSKCN1NW1P3.

Jalonick, Mary Clare. "Environmental Panels' Chairmen Chip Away at Endangered Species Act, Refocusing Resources and Definitions." *CQ Weekly*, March 27, 2004, 756.

Johnson, Steven M., "Disclosing the President's Role in Rulemaking: A Critique of the Reform Proposals." *Catholic University Law Review* 60 (2010): 1003–1044.

Just, Raymond A. "Intergenerational Standing under the Endangered Species Act: Giving Back the Right to Biodiversity after Lujan v. Defenders of Wildlife." *Tulane Law Review* 71 (1996): 597–634.

Kapucu, Naim, and Vener Garayev. "Designing, Managing, and Sustaining Functionally Collaborative Emergency Management Networks." *American Review of Public Administration* 43, no. 3 (2012): 312–330.

Kelleher, Christine A., and Susan Webb Yackee. "Who's Whispering in Your Ear? The Influence of Third Parties over State Agency Decisions." *Political Research Quarterly* 59, no. 4 (2006): 629–643.

Koebele, Elizabeth A. "Assessing Outputs, Outcomes, and Barriers in Collaborative Water Governance: A Case Study." *Journal of Contemporary Water Research and Education* 155, no. 1 (2015): 63–72.

———. "Cross-Coalition Coordination in Collaborative Environmental Governance Processes." *Policy Studies Journal* 48, no. 3 (2019): 727–753.

Koontz, Thomas M., and Elizabeth Moore Johnson. "One Size Does Not Fit All: Matching Breadth of Stakeholder Participation to Watershed Group Accomplishments." *Policy Sciences* 37, no. 2 (2004): 185–204.

Korfmacher, Katrina Smith, and Thomas M. Koontz. "Collaboration, Information, and Preservation: The Role of Expertise in Farmland and Preservation Task Forces." *Policy Sciences* 36, no. 3–4 (2003): 213–236.

Layzer, Judith A. *Open for Business: Conservatives' Opposition to Environmental Regulation.* Cambridge, MA: MIT Press, 2012.

LeLoup, Lance T. *The Fiscal Congress: Legislative Control of the Budget.* Westport, CT: Greenwood, 1980.

Lewis, David E. "The Contemporary Presidency: The Personnel Process in the Modern Presidency." *Presidential Studies Quarterly* 42, no. 3 (2012): 577–596.

———. *The Politics of Presidential Appointments: Political Control and Bureaucratic Performance.* Princeton, NJ: Princeton University Press, 2010.

Li, Ya-Wei. "Section 4(d) Rules: The Peril and the Promise." Defenders of Wildlife ESA Policy White Paper Series, 2017.

Lowande, Kenneth. "Politicization and Responsiveness in Executive Agencies." *Journal of Politics* 81, no. 1 (2019): 33–48.

Lubell, Mark. "Collaborative Environmental Institutions: All Talk and No Action?" *Journal of Policy Analysis and Management* 23, no. 3 (2004): 549–573.

———. "Collaborative Institutions, Belief-Systems, and Perceived Policy Effectiveness." *Political Research Quarterly* 56, no. 3 (2003): 309–323.

Malcom, Jacob W., and Ya-Wei Li. "Data Contradict Common Perceptions about a Controversial Provision of the US Endangered Species Act." *Proceedings of the National Academy of Sciences* 112, no. 52 (2015): 15844–15849.

Massey, Calvin R. "The Locus of Sovereignty: Judicial Review, Legislative Supremacy, and Federalism in the Constitutional Traditions of Canada and the United States." *Duke Law Journal* 1990, no. 6 (1990): 1229–1310.

May, Peter J., and Soren C. Winter. "Politicians, Managers, and Street-Level Bureaucrats: Influences on Policy Implementation." *Journal of Public Administration Research and Theory* 19, no. 3 (2009): 453–476.

Mayhew, David R. *Congress: The Electoral Connection.* New Haven, CT: Yale University Press, 1974.

McIlwain, Charles Howard. *The High Court of Parliament and Its Supremacy: An Historical Essay on the Boundaries between Legislation and Adjudication in England.* Vol. 105. New Haven, CT: Yale University Press, 1910.

Mikva, Abner J. "Congress: The Purse, the Purpose, and the Power." *Georgia Law Review* 21 (1986): 1–16.

Moe, Terry M. "Regulatory Performance and Presidential Administration." *American Journal of Political Science* (1982): 197–224.

Moore, Elizabeth A., and Tomas M Koontz. "Research Note a Typology of Collaborative Watershed Groups: Citizen-Based, Agency-Based, and Mixed Partnerships." *Society and Natural Resources* 16, no. 5 (2003): 451–460.

"No Consensus Reached on Forest Protection." *CQ Almanac* 1992, no. 48 (1993): 277–280.

Office of Congressman Raul M. Grijalva. "Chair Grijalva: President Trump's Assault on Endangered Species Will Add to Extinction Crisis as a Favor to Industry." 2019. Available at https://grijalva.house.gov/press-releases/chair-grijalva-president-trumps-assault-on-endangered-species-will-add-to-extinction-crisis-as-a-favor-to-industry/.

Olson, Erik D. "The Quiet Shift of Power: Office of Management and Budget Supervision of Environmental Protection Agency Rulemaking under Executive Order 12291." *Virginia Journal of Natural Resources Law* 4 (1984): 1–80.

O'Reilly, James T. "FDA Rulemaking after the 104th Congress: Major Rules Enter the Twilight Zone of Review." *Food and Drug Law Journal* 4 (1996): 677–696.

Parks, Julie A. "Lessons in Politics: Initial Use of the Congressional Review Act Comment." *Administrative Law Review* 1 (2003): 187–210.

Pautz, Michelle C. "Next-Generation Environmental Policy and the Implications for Environmental Inspectors: Are Fears of Regulatory Capture Warranted?" *Environmental Practice* 12, no. 3 (2010): 247–259.

———. "Trust between Regulators and the Regulated: A Case Study of Environmental Inspectors and Facility Personnel in Virginia." *Politics and Policy* 37, no. 5 (2009): 1047–1072.

Pautz, Michelle C., and Marcy H. Schnitzer. "Policymaking from Below: The Role of Environmental Inspectors and Publics." *Administrative Theory and Praxis* 30, no. 4 (2008): 450–475.

Plumer, Brad, and Coral Davenport. "Science under Attack: How Trump Is Sidelining Researchers and Their Work." *New York Times*, December 28, 2019. Available at https://nytimes.com/2019/12/28/climate/trump-administration-war-on-science.html.

Pressman, Jeffrey L., and Aaron B. Wildavsky. *Implementation: How Great Expectations in Washington Are Dashed in Oakland: Or, Why It's Amazing That Federal Programs Work at All, This Being a Saga of the Economic Development Administration as Told by Two Sympathetic Observers Who Seek to Build Morals on a Foundation of Ruined Hopes*. Berkeley: University of California Press, 1973.

Rohlf, Daniel J. "Section 4 of the Endangered Species Act: Top Ten Issues for the Next Thirty Years." *Environmental Law* 34, no. 2 (2004): 483–554.

Rosenberg, Morton. "Congressional Review of Agency Rulemaking: An Update and Assessment of the Congressional Review Act after a Decade." *Congressional Research Service, Library of Congress.* 2008.

———. "Presidential Control of Agency Rulemaking: An Analysis of Constitutional Issues That May Be Raised by Executive Order 12,291," *Arizona Law Review* 23 (1981): 1199–1234.

———. "Whatever Happened to Congressional Review of Agency Rulemaking: A Brief Overview, Assessment, and Proposal for Reform." *Administrative Law Review* 4 (1999): 1051–1092.

Sabatier, Paul A., Will Focht, Mark Lubell, Zev Trachtenberg, Arnold Vedlitz, Marty Matlock, Michael E Kraft, and Sheldon Kamieniecki. *Swimming Upstream: Collaborative Approaches to Watershed Management*. Cambridge, MA: MIT Press, 2005.

Shane, Peter M. "Presidential Regulatory Oversight and the Separation of Powers: The Constitutionality of Executive Order No. 12291." *Arizona Law Review* 23 (1981): 1235–1266.

Sheikh, P. A., E. H. Ward, and R. E. Crafton. *Final Rules Changing Endangered Species Act Regulations*. Edited by Congressional Research Service. Washington, DC, 2019.

Sinclair, Ward. "'Pork Panic' Sweeping Congress in Wake of Darter's Rescue." *Washington Post*, June 28, 1978.

Smith, Douglas W., Wayne G. Brewster, and Edward E. Bangs. "Wolves in the Greater Yellowstone Ecosystem: Restoration of a Top Carnivore in a Complex Management Environment." In *Carnivores in Ecosystems: The Yellowstone Experience*, edited by Tim W. Clark, A. Peyton Curlee, Steven C. Minta, and Peter M. Karieva, 103–126. New Haven, CT: Yale University Press, 1999.

Stahl, Peter D., and Michael F. Curran. "Collaborative Efforts towards Ecological Habitat Restoration of a Threatened Species, Greater Sage-Grouse, in Wyoming, USA." *Land Reclamation in Ecological Fragile Areas—Proceedings of the 2nd International Symposium on Land Reclamation and Ecological Restoration*. LRER 2017, 251–254. Available at https://doi.org/10.1201/9781315166582-53.

Stearns, Maxwell L. "From Lujan to Laidlaw: A Preliminary Model of Environmental Standing Citizen Suits and the Future of Standing in the 21st Century: From Lujan to Laidlaw and Beyond." *Duke Environmental Law and Policy Forum* 2 (2000): 321–388.

Steeves, Michael. "The EPA's Proposed CAFO Regulations Fall Short of Ensuring the Integrity of Our Nation's Waters." *Journal of Land Resources and Environmental Law* 22 (2002): 367–397.

Taylor, Phil, and Jean Chemnick. "House Panel Advances Rider-Laden Interior, EPA Bill." *Greenwire*, July 7, 2011.

Ulibarri, Nicola. "Tracing Process to Performance of Collaborative Governance: A Comparative Case Study of Federal Hydropower Licensing." *Policy Studies Journal* 43, no. 2 (2015): 283–308.

U.S. Department of the Interior. "Trump Administration Improves the Implementing Regulations of the Endangered Species Act. 2019. Available at https://www.doi.gov/pressreleases/endangered-species-act.

U.S. Fish and Wildlife Service. "Mission." 2020. Available at https://www.fws.gov/le/.

——, ed. 1987. *Northern Rocky Mountain Wolf Recovery Plan.* Denver, CO. 1987.

U.S. Fish and Wildlife Service, Interior; National Marine Fisheries Service, National Oceanic and Atmospheric Administration, Commerce, Proclamation. "Endangered and Threatened Wildlife and Plants; Regulations for Interagency Cooperation." *Federal Register* 84, no. 166 (August 27, 2019): 44976–45016. Available at https://www.govinfo.gov/content/pkg/FR-2019-08-27/pdf/2019-17517.pdf.

U.S. Fish and Wildlife Service and National Marine Fisheries Service. Post-Delisting Monitoring Plan Guidance under the Endangered Species Act. 2018.

Vig, Norman, and Michael E. Kraft. *Environmental Policy: New Directions for the Twenty-First Century.* Washington, DC: CQ Press, 2018.

Wehner, Joachim. "Assessing the Power of the Purse: An Index of Legislative Budget Institutions." *Political Studies* 54, no. 4 (2006): 767–785.

Weinberg, Philip. "Are Standing Requirements Becoming a Great Barrier Reef against Environmental Actions?" *NYU Environmental Law Journal* 7 (1999): 1–26.

West, William F., and Connor Raso. "Who Shapes the Rulemaking Agenda? Implications for Bureaucratic Responsiveness and Bureaucratic Control." *Journal of Public Administration Research and Theory* 23, no. 3 (2013): 495–519.

Wood, B. Dan. "Modeling Federal Implementation as a System: The Clean Air Case." *American Journal of Political Science* 36, no. 1 (1992): 40–67.

Yackee, Jason Webb, and Susan Webb Yackee. "A Bias towards Business? Assessing Interest Group Influence on the U.S. Bureaucracy." *Journal of Politics* 68 (2006): 128–139.

Yackee, Susan Webb. "Sweet-Talking the Fourth Branch: The Influence of Interest Group Comments on Federal Agency Rulemaking." *Journal of Public Administration Research and Theory* 16 (2006):103–124.

Younger, Cally, and Sam Eaton. "Lessons Learned from the Greater Sage-Grouse Land Use Planning Effort." *Idaho Law Review* 53, no. 2 (2018): 373–384.

5

What Happens Next?

Insights from Environmental Inspectors

MICHELLE C. PAUTZ

After environmental rules and regulations are established, it is easy to assume that the difficult part is over, that it is simple to check on and enforce those rules and regulations. Perhaps even mental pictures emerge of environmental cops with badges and green flashing lights protecting the environment. Since images abound of government officials wielding their power and ability to police behavior, it seems logical to assume the same exists in the environmental context. It would seem inspectors must just be another category of zealous government officials who wield significant power. But, as the preceding chapters have demonstrated, it is much more complicated when it comes to the work of environmental inspectors—and there are no green flashing lights. Different types of government officials have distinct kinds of power, as described in Chapter 1, and environmental inspectors have a limited range of options to compel compliance with regulations.

Much of the focus of this volume is on how implementation of environmental policy works and the stories of those individuals on the front lines of environmental protection day in and day out. This chapter focuses on what happens after rules and regulations are promulgated, beginning with a discussion of inspectors themselves and some of their own stories. The roles of inspectors are varied and affected by the internal and external contexts of their agencies. These factors exert influence over how inspectors interact with their regulatory counterparts in firms and the nature of those interactions. Woven together throughout this chapter are examples from inspectors alongside discussions of academic investigations about these frontline workers, regulatory enforcement approaches and styles, and the nature of regulatory interactions. By the conclusion of the chapter, the essential role of environmental inspectors will be evident, and it is hoped that readers will gain an understanding of and appreciation for the complexities associated with inspectors' positions in pursuing environmental protection.

Case Study: Air Inspectors in the Midwest

As a starting point for understanding the work of inspectors, consider the case of an air quality inspector from the Midwest. Brenda[1] started working with environmental regulations as an intern while she was completing her undergraduate degree in environmental science. Upon graduating, she went to work for the same government agency, serving in a variety of roles, and she has been an inspector for ten years. She is responsible for inspecting about four dozen facilities in her assigned geographic area of the state and conducts between twelve and fifteen inspections annually. In addition to conducting physical site inspections, Brenda spends much of her time working with facilities on their reporting requirements and other compliance obligations.

The state where Brenda works essentially took a federal environmental rule for a particular hazardous air pollutant, codified it in state rules, and substituted volatile organic compounds (VOCs) for the pollutant so the state did not have to rewrite the rule—that is her take on it anyway. The problem, as Brenda sees it, is that VOCs require different controls than hazardous air pollutants do, but she is not the rule writer, she is just the inspector. The rule is not working, but Brenda has to enforce it. One of the fiberglass manufacturers she is responsible for is having a hard time with the VOC rule and figuring out how to track its VOC emissions. Part of the challenge is determining how to track the information in a way that makes sense. But Brenda cannot give the facility tools to track the emissions, because that violates her role. Ultimately, her goal is compliance with environmental rules, so Brenda wants to help the facility while being mindful of her role as a regulator—and the facility wants help, too. Therefore, she shows the manufacturer how a spreadsheet could be set up for a fictional company to track VOC emissions, thereby leaving it up to the facility to do as it deems appropriate. In the end, the fiberglass facility, which was not in compliance with some of the reporting requirements, is now in compliance because Brenda understood that the rule was hard to follow and she was willing to take the time to help the facility make sense of the rule. Brenda understands her role is monitoring and ensuring facilities comply with air regulations to protect the environment; she is not an environmental cop who issues citations for every issue of noncompliance she encounters.

This introduction about Brenda's experiences helps convey inspectors' role as an intermediary between rule writers at both the federal and state levels and the firms that are regulated by those rules. The work that Brenda and other front-line regulators do every day constitutes the implementation of environmental policy, as translated into rules and regulations, and their work is essential to protecting the environment.

Environmental Inspectors: Who They Are and What They Do

The preceding case provides a glimpse into who inspectors are and what they do to pursue environmental protection, but it is just one example. It may already be apparent from Brenda's story that the work of an environmental inspector is rarely monotonous and involves plenty of work both inside and outside the office, since inspectors are the bridge between environmental rules and implementation. This positionality makes them street-level bureaucrats or front-line workers, because they "deal with people who may not voluntarily desire their services and who experience some ambiguity in exercising appropriate control."[2]

Environmental inspectors are employed by federal, state, and, in some cases, local environmental protection agencies. These agencies include, for example, the U.S. Environmental Protection Agency, the Virginia Department of Environmental Quality, and the Division of

Air Quality in the Cleveland Department of Public Health. All fifty states and the District of Columbia have one or more agencies whose mission is environmental protection, and some states, such as Ohio, have local agencies as well. Traditionally in the United States, environmental inspectors specialize in a particular area or environmental media, such as air quality, solid waste, hazardous waste, drinking water, or surface water. And within these broad categories, there may be more specific delineations, such as asbestos inspectors or concentrated animal feeding operations inspectors. These areas of specialization are a function of the regulatory structure that is devised around major statutes, such as the Clean Air Act or the Safe Drinking Water Act. In these agencies, inspectors are often at the base of the organizational pyramid. Some inspectors make this work the centerpiece of their professional lives, while others gain experience as an inspector and move into management and leadership of the environmental protection agency or pursue other career options.

It is difficult to make broad generalizations about inspectors as individuals since they serve in a variety of agencies at all levels of government. Michelle Pautz and Sara Rinfret conducted one of the few larger-scale studies about environmental inspectors and gathered data from more than twelve hundred state-level environmental inspectors. They found that most inspectors are men (68 percent), with nearly half (43 percent) holding bachelor's degrees in various fields from biology to environmental engineering.[3] It is worth noting that the educational backgrounds and fields of study among inspectors do vary quite a bit, so it is not simply environmental scientists and engineers who find their way into these roles but also economists and political scientists alongside individuals with many other backgrounds. Commensurate with other studies of civil servants, this research also found the majority of inspectors remain in their roles for ten or more years.[4]

This discussion helps provide a picture of who environmental inspectors are, thereby providing a foundation to explore what they do. The terms *inspector* and *regulator* might convey that the person in this position conducts routine visits to a facility to ensure compliance. Although that is part of the job, it is only one aspect.[5] In the course of any given day, inspectors regularly interact with members of the regulated community to ensure compliance with environmental rules.

This might entail physical site inspections of a facility, but it may also include review of monitoring reports and other data gathered from and submitted by regulated facilities, following up on compliance issues, and engaging enforcement proceedings for a facility that is in violation of regulations. The work might also include investigating complaints about a facility and interacting with the public when questions or concerns are raised, such as noxious smells from a manufacturing plant. Inspectors are responsible for staying up to date on the latest rules or modifications of existing rules from federal and state agencies.

All these activities may be even more daunting after considering the kinds of facilities an inspector may be assigned. An air quality inspector might be responsible for monitoring compliance at printing facilities, plastics manufacturers, power plants, and dry cleaners—and might have to monitor more than 150 facilities.[6] This means the inspector has to keep up with not only hundreds of facilities but also the different kinds of operations and production processes in each facility, not to mention the latest technological advances in each industry that could affect air pollution. A hazardous waste inspector may be responsible for only a handful of hazardous waste facilities, but the monitoring and other regulatory obligations for these kinds of facilities can be extremely complex. This scope of responsibility can be daunting, but it is simply part of the job of an inspector.

Even though inspectors' tasks are perhaps more varied than might initially be expected, it is worth exploring the particulars of compliance inspections. Simply put, the point of an inspection is to determine whether a firm is in compliance with applicable regulations; this can automatically make interactions between inspectors and individuals at a regulated firm confrontational. To prepare for an inspection, an inspector will review a facility's permit, its compliance history, and the applicable regulations the facility is subject to. Some facilities will be inspected multiple times a year; in other instances, a facility might not be inspected more than once every several years. Inspection frequency is dictated by the size of the facility, why it is regulated, and its compliance history.

Inspections can be rather brief or span multiple days. Upon arriving—usually unannounced—at the facility, an inspector will begin by meeting with facility officials and discussing the purpose of the visit. Typically, an inspector will review monitoring logs and other documentation required by regulations and tour the facility to see firsthand processes and practices in place that are required by the regulations. Over the course of the visit, the inspector will engage with her regulatory counterpart at the facility. Depending on the size and nature of the facility, this might be a dedicated environmental compliance manager or the owner and operator of the facility fulfilling many roles. At the conclusion of the inspection, the inspector will review the inspection, answer questions, and let the facility officials know about any likely follow-up actions. Inspectors usually submit formal, written inspection reports, but many inspectors prefer to give their regulatory counterparts an informal sense of the probable outcome at the close of an inspection.[7]

During an inspection, an inspector is likely to encounter all sorts of circumstances, ranging from the pleasant to the not-so-pleasant. Generally speaking, inspectors expect to find that the regulated community is doing what it is supposed to be doing and is abiding by regulations.[8] Violations can run the gamut from missing copies of monitoring logs to blatant and intentional dumping of pollutants or other contaminants (though the latter is uncommon). Typically, inspectors endeavor to remedy violations as quickly as possible, because the goal is environmental protection. During a physical inspection, an inspector might point out an issue and try to convince the facility to fix it immediately before rising to the level of any formal finding of noncompliance. Issues might still be documented in an inspection report, but not in the sense of a formal compliance violation.

After an inspection, the inspector will brief her supervisor on any issues noted and recommend any next steps. If a violation is noted, the first formal step might be the notification of a finding of noncompliance and a request for corrective action—a formal request to fix the problem, usually within a specific time frame (e.g., sixty days). After the request has been issued, the inspector follows up with the facility and hopes to discover that the problem has been solved. If a request for corrective action does not bring about the desired outcome, a warning letter may be drafted by the agency, representing the next step in a potential enforcement process. These formal letters requesting a facility address the noncompliance often become public record. If this escalation does not solve the issue, the regulatory agency can move for a notice of violation, which usually involves attorneys from both the regulatory agency and the facility. Notices of violation can prompt legal battles to pursue regulatory compliance, and they represent significant escalation in the compliance process. These sorts of scenarios are infrequent, but they can and do happen. The involvement of the inspector gives way to high-ranking officials in environmental protection agencies and attorneys representing all parties. Although these scenarios involving different types of action provide a sense of how the enforcement process may unfold, they do not capture the complex situations an inspector will face.

Sometimes an inspector finds a facility is not in compliance with applicable regulations, and it is not a simple task to determine the next course of action. For example, if an air inspector, like Brenda, is walking the production floor of a facility and notices that drums of solvent-soaked rags are uncovered, then the facility is violating rules because the drums should be covered to contain air pollutants. However, the inspector must decide how to respond and determine the significance of the issue. An inspector may decide that simply pointing out the uncovered drums to the facility representative and gauging that representative's reaction is sufficient to determine next steps.

Perhaps the facility representative immediately recognizes the problem, is clearly aware there is a violation, and takes rapid and decisive steps to rectify the situation. This is probably exactly what the inspector hopes for and simply makes a note of it in the inspection report. Alternatively, the inspector may point out the uncovered drums, and the facility representative looks bewildered and shrugs his shoulders. In this scenario, the inspector will likely respond differently. Another possibility is that the inspector chooses to say nothing about the issue, documents the problem, and then recommends to her supervisor that the facility be formally notified it is not in compliance. Many of the intertwined factors that help explain and predict how an inspector will respond are explored throughout this chapter. As Brenda's story demonstrates, the work of inspectors has evolved such that inspectors are "no longer to merely inspect and enforce, but also to educate, to negotiate, and to make compromises."[9]

While the focus here is on the perspective of environmental inspectors, it is important to acknowledge the kaleidoscope of motivations from the regulated community, because those motives help us understand the wide variety of situations inspectors encounter. Christine Parker and Vibeke Lehmann Nielsen summarize compliance motivations in three categories: economic, social, and normative.[10] Economic motives for firms involve calculating the costs of compliance versus the costs of noncompliance and the effect of those decisions on the profitability of the firm. Social motives speak to what key stakeholders—shareholders, customers, members of the public—may think (and likely do) based on a firm's regulatory behavior. Put differently, if a firm decides that an important piece of its image and reputation is environmental stewardship, then compliance with regulations and promoting the environment is integral to the business operations of the firm. Any failure to live up to those expectations could be detrimental to the firm's reputation—and profitability. Normative motivations are often more abstract and encompass the firm's desire to behave in an ethical way, regardless of the economic costs or what stakeholders and customers deem appropriate. As might be expected, firms can exhibit varying combinations of these motives, and these motives undoubtedly affect their interactions with inspectors.

Given the essential role of inspectors in implementing environmental regulation, it is perhaps a bit surprising they are often overlooked in both environmental policy literature and in regulatory governance literature—although regulators more broadly are duly acknowledged in the latter. Some of the earliest efforts to understand inspectors came from Eugene Bardach and Robert Kagan in *Going by the Book* and Keith Hawkins in *Environment and Enforcement*.[11] In discussing regulatory challenges, Bardach and Kagan call attention to the difficult, but vital, work of inspectors, and the authors lay the foundation for discussions of the importance of discretion and cooperation in inspectors' work. These efforts coincide with the emergence of the street-level bureaucrat or front-line worker literature that emerged thanks to Michael Lipsky and Steven Maynard-Moody and Michael Musheno that focused on individuals such as teachers, social workers, and police officers and their day-to-day work life.[12] These areas of research inform one another in recent work on inspectors and other front-line regulators, including Stephen Van de Walle and Nadine Raaphorst's *Inspectors and Enforcement at the Front Line of Government*.[13]

With this understanding of inspectors and their roles, attention shifts to their work environment to explain the influences on inspectors as they approach their interactions with the regulated community.

The Work Environment of Inspectors

The work of inspectors is fundamentally social because it involves interactions with not only individuals at regulated firms but also individuals in the inspector's own organization as well as other environmental protection officials at different levels of government and in the community. As Van de Walle and Rapphorst note, "Inspectors' discretion is not random but socially constrained by their organization's culture, their training, their colleagues at work, or their resources."[14] Therefore, it is essential to consider the organizational structure and culture inspectors work in and public perceptions of their work.

The structural realities of inspectors' work environment are complex to say the least. Recall that environmental inspectors work for local, state, or federal government agencies. Despite all these agencies working to pursue environmental goals, significant differences are likely in various agencies' approaches and priorities; in some cases, the agencies might have competing regulatory requirements. As noted throughout this volume, the Environmental Protection Agency establishes much of the environmental regulatory framework in the United States, but states can and do go beyond the federal guidelines. Sometimes those differences cause tension and even result in court battles, as differences in auto emissions regulations in California have demonstrated. Members of the regulated community may interact with inspectors from each of these agencies—and perhaps even multiple inspectors from each agency, as many agencies have different divisions dedicated to the various environmental media. In other words, Brenda might be the air quality inspector a facility deals with, but that same facility might also deal with Harold the waste inspector, Beth from water, and a federal EPA inspector from time to time. With many inspectors in many agencies, differences are likely to emerge.

Moreover, each government entity has its own organizational culture, which means that the agency has its own values, norms, beliefs, and attitudes about its purpose and how it performs its duties. All organizations develop their own cultures, and that affects the work they do and how they are perceived by their stakeholders. Returning to the example, Brenda and Harold might both work for the same state agency that embraces an attitude of doing right by the environment without jeopardizing much-needed economic development in their state, whereas the EPA inspector might come from a culture that prioritizes strict adherence to regulations or collecting revenue from fines.

It is worth noting that perceptions about regulators, and in particular environmental inspectors, also play a role in the work environment. Broadly speaking, Americans prefer to keep government involvement in their lives to a minimum and are rarely excited by anyone from the government that appears to meddle. Charles Goodsell in *The New Case for Bureaucracy* aptly describes the negative attitudes citizens often have about government and the adverse impacts these views can have on government work.[15] These negative attitudes extend to environmental inspectors.

Unlike in other regulatory contexts, such as workplace safety regulations, environmental inspectors typically cannot walk into a facility and shut down operations. This difference in regulatory contexts, however, is not widely known, so fear and trepidation are among some of the tamer responses that members of the regulated community might demonstrate toward inspectors. In other instances, inspectors might encounter include outright loathing and disdain from the regulated community. These perceptions that are held about inspectors are important in considering the everyday environment in which inspectors perform their jobs. This environment has an impact on the information shared with inspectors and how they interact with members of the regulated community and inspectors in other divisions or agencies.

Regulatory Interactions

The nature of the interactions between inspectors and the regulated community often varies quite a bit from the perceptions of those interactions—after all, assumptions that inspectors wield green flashing lights seem plausible. Undoubtedly, the interactions inspectors have with members of the regulated community are at the core of inspectors' work, yet less is understood about these interactions than might be expected. As discussed previously, Bardach and Kagan's work was among the first to call attention to the role of environmental inspectors and the challenges they confront; since then, scholars have investigated and attempted to unpack these interactions. Frequently discussions of these interactions center on enforcement strategies and styles. Although these terms are often conflated, an important distinction relates to the individual versus the collective.[16] An agency's approach to interacting with the regulatory community is encapsulated by the term *enforcement strategy*, while the term *enforcement style* refers to the approach of the individual inspector. Peter May and Robert Wood summarize enforcement styles as "the character of the day-to-day interactions of inspectors when dealing with representatives of regulated entities."[17] And it is worth pointing out that both styles and strategies "are neither mutually exclusive nor independent. Regulatory agencies and their inspectors may adopt them individually or in combination."[18]

Briefly, enforcement strategies are established by the agencies that environmental inspectors work for; the strategies dictate the general attitude around regulatory compliance and usually establish agency priorities. For example, an environmental agency may focus on a particular industry sector or establish guidelines for inspectors to follow that concern the frequency of certain types of inspections. These decisions may be a function of an agency's leadership, resources, and perhaps even the political context in which the agency operates. Some research explores enforcement strategies,[19] but the focus here is on the actions of individual inspectors—therefore, the interest is on enforcement styles.

Regulatory Enforcement Styles

A considerable body of research exists on the enforcement styles of inspectors in a variety of regulatory contexts.[20] "Enforcement style can be defined as the general approach adopted by an inspector in the course of performing his regulatory duties," according to Carlos Lo, Ning Liu, and Pansy Li.[21] In other words, enforcement style describes how an inspector approaches her work with members of the regulatory community. This encapsulates the inspector's day-to-day behavior in her work with the regulated community. Initial research on enforcement styles advanced the idea that inspectors adopt one of two styles: a formal, rules-oriented approach or a flexible, outcome-oriented approach.[22] The formal, rules-oriented approach can be summarized as an enforcement style that is a strict, follow-the-rules-exactly approach. The emphasis with this style is on consistency and applying the rules uniformly regardless of the circumstances. Inspectors embracing this one-size-fits-all approach believe that being consistent is the way to ensure compliance among the regulated community and that any flexibility leaves the regulatory regime open to manipulation.

In contrast, inspectors who adopt a flexible, results-oriented approach seek compliance with regulations through cooperative and collaborative interactions with the regulated community. This is not to say this approach enables the regulated community to get away with not complying; rather, the focus is on the outcome instead of strict adherence to the regulation. This approach recognizes the context within which a regulated facility operates along with other factors that may influence the firm's ability to comply and the behavior of the individuals at that firm. Consequently, an inspector is in pursuit of a regulatory goal—such as minimal amounts of hazardous air pollutants that are emitted from a smokestack—rather than rigid conformity with a particular rule. Again, this does not mean that inspectors allow firms to break rules. It conveys that inspectors want to work with the regulated community to achieve regulatory goals and do not see their role in an adversarial, us-versus-them manner. Instead of an emphasis on a universal approach, these inspectors believe cooperation is the best way forward.

While the initial understandings of regulatory enforcement styles were predicated on the notion that inspectors embraced one of these two styles, it quickly became evident that such a binary approach was too simplistic. Research has subsequently demonstrated that these two approaches should be thought of as two ends of a regulatory enforcement style spectrum and that inspectors embrace in varying combinations attributes of each style.[23] One inspector quoted in Pautz and Rinfret's study of state environmental regulators sums up the work of the inspector as follows: "Being a regulator is more than enforcing the rules. A regulator needs to understand the purpose and intent of the rules and regulations and also have an understanding of the process and objectives of the regulated community. It is important to be able to explain to the regulated community what rules and regulations apply to them, why they apply, and then be able to have an educated conversation regarding achieving and maintaining compliance with the rules and regulations."[24]

Further research has asserted five types of enforcement styles along the continuum: formalism, coercion, education, prioritization, and accommodation.[25] Formalism encapsulates the legal basis for the regulation and the use of legal tools in enforcement actions, leading to a highly rigid, rule-driven style. Moving along the style spectrum, coercion describes a style that emphasizes sanctions for noncompliant behavior guided by the belief that only threats will compel compliance. Toward the middle of the spectrum is the education style, which seeks compliance through communication and education about the regulations. Inspectors who practice this style believe that compliance can be achieved by helping the regulated community understand the rules.

The prioritization style focuses on the attainment of a particular goal or outcome rather than on how that outcome is achieved; in other words, inspectors with this enforcement style prioritize based on the most significant (in this case, environmental) concerns and focus on achieving a public good. Finally, accommodation is the enforcement style that accounts for the external forces—ranging from political influence to the regulated community—that drive how inspectors pursue compliance and enforcement and efforts to reconcile those competing demands from stakeholders. It is important to note that none of these five styles is distinct, and they can easily blend with other styles. Inspectors may embrace many of these styles, and that combination could vary depending on the regulated firm. Finally, all enforcement styles have their strengths and weaknesses and one style should not be lauded as the better approach over another.

In addition to research that describes how inspectors pursue compliance among the regulated community, efforts have been undertaken to explore the factors that influence the enforcement style of a particular inspector. Lo, Liu, and Li summarize that enforcement styles are determined by (1) societal factors, (2) organizational factors, and (3) individual factors.[26] Societal factors refer to the forces at work outside the inspectors' agency that range from governmental pressure or support—perhaps political influence—and pressure and support from the public at large. Put differently, societal factors encapsulate the attitudes and preferences of the whole of government and society about regulation and compliance in a particular regulatory regime.

Organizational factors are internal to the inspectors' agency and might include the support from management and the agency around particular goals or actions. Also internal to the organization is the ambiguity surrounding the implementation and administration of the regulations. The environment inspectors encounter in their agency may support discretion and flexibility in pursuing compliance goals, or there may be little room for inspectors to exercise discretion in their work; therefore, the level of ambiguity can be a significant factor. Finally, inspectors' individual traits—demographics, past experiences, and even their values and commitment to the regulatory regime and its goals—are likely to influence their enforcement style.

By way of summary, inspectors are not a monolith acting as regulatory automatons; rather, they are individuals who are influenced by a wide range of factors—and to varying degrees by those factors—that determine how they approach pursuing compliance and working with the regulated community. Inspectors' enforcement styles are not universal and are likely to vary both in regard to the specific regulated firm and over time. At first glance, this may cause concern, but a closer consideration reveals it is highly unlikely that rules and procedures can be dictated for every situation an inspector encounters. Accordingly, it is more advantageous to ensure that inspectors have the administrative discretion needed to deal with a given situation. Because inspectors exercise discretion in how they perform their work, the enforcement styles they employ can have various effects on the regulated community and overall regulatory goals.

As might be expected, an inspector's enforcement style is likely to affect compliance by the regulated firm. Peter May and Søren Winter detail the possible effects of an inspector's enforcement style on compliance; they note that measuring the effects is extremely difficult, and "perhaps the most important thing to note about the effects of enforcement style on compliance is that they are not uniform."[27] Additionally, May and Winter discuss other effects, including the regulated firm's degree of awareness of the rules and regulations along with its willingness to cooperate with the inspector. Moreover, these researchers and others find that a more cooperative or facilitative regulatory enforcement style may enhance perceptions of cooperation between the inspector and the regulated community, whereas a more formalistic approach undermines that interaction.[28]

Social Nature of Regulatory Compliance

It is important to recognize the importance of regulatory enforcement styles and the reality that they are not neatly encapsulated into two well-defined styles; inspectors use varying combinations of approaches depending on the situation. Equally important is understanding that the regulated firm demonstrates the inherently social nature of regulatory implementation and enforcement. In other words, regulations are not as simple as what is written on the pages of the *Federal Register* or in similar state documents, for instance. Instead, regulatory compliance is socially constructed through the interactions between the regulators and the regulated; compliance is fundamentally a social process.[29] Robert Kagan, Neil Gunningham, and Dorothy Thornton articulate the social nature of compliance as follows:

> Environmental regulation, for example depends almost entirely on regulated business firms to devise, finance and operate the technologies that prevent, measure or treat pollution. Inside regulated companies the day-to-day effectiveness of many regulatory compliance measures depends on the capacity and diligence of the corporate employees assigned to maintain the equipment, monitor quality control systems, train operators and take appropriate action when problems occur. In sum, effective regulation requires imaginative cooperation as much or even more than it requires government monitoring and legal coercion.[30]

Ultimately, the interactions between inspectors and the regulated community convey the social nature of the implementation and enforcement of regulations. And the approaches and styles employed by individuals and their organizations demonstrate that they must work together, and they must interact as they construct regulatory processes; indeed, neither party can be solely responsible for compliance. Regulatory compliance depends on the exchange of information between regulators and members of the regulated community because of high levels of information asymmetry. Regulators are not at a facility every day monitoring compliance; they rely on the information provided by the regulated firm and have to presume that the data submitted are accurate. From the perspective of the regulated community, inspectors might need assistance understanding the text of the regulation and the guidance issued from the regulatory agencies on how to comply with the regulation. As has been demonstrated here—and throughout this volume—regulations are complex and not necessarily straightforward. The interdependence and social nature of regulatory interactions often stand in contrast to perceptions of environmental cops policing the negligent regulated community. Although such perceptions may still be deeply rooted for some observers, research demonstrates that the reality of regulatory compliance is quite a bit different: most regulated firms desire to be in compliance with regulations.[31] Accordingly, regulatory governance scholars note the need for different frameworks for considering regulatory compliance, and the notion of responsive regulation has been widely embraced.

Responsive Regulation

Ian Ayers and John Braithwaite's 1992 volume *Responsive Regulation: Transcending the Deregulation Debate* compelled a fundamental shift in how scholars investigate regulatory interactions. Briefly put, "Responsive regulation emphasizes cooperation, trust, and dialogue between parties rather than strict top-down enforcement, and encourages inspectors to adapt to inspectees' behavior during inspector-inspectee encounters when making decisions."[32] The hallmarks of responsive regulation include pursuit of compliance by working with—rather than against—the regulated community and responding to firms based on their interactions with inspectors. As a result, the responses to a particular firm may vary depending on the firm's compliance record and the nature of the interactions with the firm.

Ayers and Braithwaite crafted an "enforcement pyramid" to help an inspector respond to and work with the regulated community to secure compliance and adjust enforcement approaches based on their responses. The inspector responses range from strict, universal punishment, to enforced self-regulation, to self-regulation.[33] This approach moves away from a punitive, one-size-fits-all approach to regulatory compliance. The earlier example of the inspector finding uncovered drums of solvent-soaked rags and convincing the firm to fix the problem immediately without initiating a formal enforcement proceeding illustrates responsive regulation in action. In the three decades since responsive regulation was introduced, it has garnered a wealth of research and seen growth and refinement of its initial ideas. A thorough review of that scholarship is beyond the scope of this chapter, but responsive regulation is introduced here to demonstrate that regulatory theory represents a clear departure from commonly held expectations about the nature of regulation. The adversarial nature of regulation that is often perceived is not the norm in both the practice and study of environmental regulation.

Concluding Thoughts

This chapter has picked up the conversation about how environmental regulations are implemented after they are promulgated, and it has demonstrated that the interactions between regulators and members of the regulated community do not involve green flashing lights and environmental police badges. Additionally, compliance with environmental regulations is usually not about forcing recalcitrant firms to comply with regulations that they are trying to circumvent. Instead this chapter illuminates the role of environmental inspectors and places their work in the broader context of the environmental regulatory regime. A discussion of the organizational environment describes the setting in which inspectors conduct their work and how they interact with members of the regulated community. A discussion of the nature of the interactions with the regulated community included an exploration of regulatory styles and strategies.

From this discussion, a few conclusions can be drawn. First, the work of environmental inspectors is critical in implementing environmental rules and regulations, and that work is far from a simplistic monitoring of what a firm is or is not doing. Compliance can be extremely complicated because environmental regulations can be extremely complicated. And compliance is inherently a social process. Second, the interactions between inspectors and members of the regulated community are essential to understanding the day-to-day realities of environmental regulation, and these interactions are far more cooperative and collaborative than might be presumed. As Pautz and Rinfret, along with others, conclude, such interactions are often amicable and contribute to positive environmental outcomes. In sum, environmental inspectors play an integral role in protecting the natural environment, and to be successful, they have to work productively with members of the regulated community. The next chapter continues this exploration by chronicling the implementation of oil and gas regulations in Colorado.

Notes

1. The identity of Brenda and the agency she works for has been kept confidential. Her stories were gathered during semistructured interviews conducted by the author in 2014.

2. Stephen Fineman, "Street-Level Bureaucrats and the Social Construction of Environmental Control," *Organization Studies* 19, no. 6 (1998): 953. See also Michael Lipsky, *Street-Level Bureaucracy: Dilemmas of the Individual in Public Services* (New York: Russell Sage Foundation, 1980); Steven Maynard-Moody and Michael Musheno, *Cops, Teachers, Counselors: Stories from the Front Lines of Public Service* (Ann Arbor: University of Michigan Press, 2003); Norma M. Riccucci, *How Management Matters: Street-Level Bureaucrats and Welfare Reform* (Washington, DC: Georgetown University Press, 2005).

3. Michelle C. Pautz and Sara R. Rinfret, *The Lilliputians of Environmental Regulation: The Perspective of State Regulators* (New York: Routledge, 2013), 18–19.

4. Ibid., 19.

5. See, for example, Pautz and Rinfret, *The Lilliputians*, 22; see also Bridget M. Hutter, *Compliance: Regulation and Environment* (Oxford: Clarendon Press, 1997).

6. Pautz and Rinfret, *The Lilliputians*, 23.

7. Ibid., 25.

8. See, for example, Pautz and Rinfret, *The Lilliputians*; Michelle C. Pautz and Sara R. Rinfret, "State Environmental Regulators: Perspectives of Trust with Their Regulatory Counterparts," *Journal of Public Affairs* 16, no. 1 (February 2016): 2838; Michelle C. Pautz, "Front-Line Regulators and Their Approach to Environmental Regulation in Southwest Ohio," *Review of Policy Research* 27, no. 6 (November 2010): 761–780; Michelle C. Pautz, "Trust between Regulators and the Regulated: A Case Study of Environmental Inspectors and Facility Personnel in Virginia, *Politics and Policy* 37, no. 5 (October 2009): 1047–1072.

9. Steven Van de Walle and Nadine Raaphorst, "Introduction: The Social Dynamics of Daily Inspection Work," in *Inspectors and Enforcement at the Front Line of Government*, ed. Steven Van de Walle and Nadine Raaphorst (New York: Palgrave Macmillan, 2019), 2–3.

10. Christine Parker and Vibeke Lehmann Nielsen, eds., *Explaining Compliance: Business Responses to Regulation* (Northampton, MA: Edward Elgar, 2011).

11. Eugene Bardach and Robert A. Kagan, *Going by the Book: The Problem of Regulatory Unreasonableness* (New Brunswick, NJ: Transaction, 1982/2003);

Keith Hawkins, *Environment and Enforcement: Regulation and the Social Definition of Pollution* (Oxford: Clarendon Press, 1984).

12. Michael Lipsky, *Street-Level Bureaucracy: Dilemmas of the Individual in Public Services* (New York: Russell Sage Foundation, 1980); Steven Maynard-Moody and Michael Musheno, *Cops, Teachers, Counselors: Stories from the Front Lines of Public Service* (Ann Arbor: University of Michigan Press, 2003).

13. Steven Van de Walle and Nadine Raaphorst, eds., *Inspectors and Enforcement at the Front Line of Government* (New York: Palgrave Macmillan, 2019).

14. Van de Walle and Raaphorst, "Introduction."

15. Charles T. Goodsell, *The New Case for Bureaucracy* (Thousand Oaks, CA: Sage/CQ Press, 2014).

16. Peter J. May and Søren C. Winter, "Regulatory Enforcement Styles and Compliance," in *Explaining Compliance: Business Responses to Regulation*, ed. Christine Parker and Vibeke Lehmann Nielsen (Northampton, MA: Edward Elgar, 2011), 222–244. See also Peter J. May and Søren Winter, "Reconsidering Styles of Regulatory Enforcement: Patterns in Danish Agro-Environmental Inspection," *Law and Policy* 22, no. 2 (April 2000): 143–173; Pautz and Rinfret, *The Lilliputians*; Carolos Lo, Ning Liu, and Pansy Li, "Dynamics of Inspectors' Enforcement Styles," in *Inspectors and Enforcement at the Front Line of Government*, ed. Steven Van de Walle and Nadine Raaphorst (New York: Palgrave, 2019), 95–123.

17. Peter J. May and Robert S. Wood, "At the Regulatory Front Lines: Inspectors' Enforcement Styles and Regulatory Compliance," *Journal of Public Administration Research and Theory* 13, no. 2 (2003): 119.

18. Lo, Liu, and Li, "Dynamics of Inspectors'," 102.

19. See, for example, John T. Scholz, "Managing Regulatory Enforcement," in *Handbook of Regulation and Administrative Law*, ed. David Rosenbloom and Richard D. Schwartz (New York: Marcel Decker, 1994), 423–463; Albert Reiss, "Selecting Strategies of Social Control over Organizational Life," in *Enforcing Regulation*, ed. Keith Hawkins and John Thomas (Boston: Kluwer-Nijhoff, 1984), 23–25; John Braithwaite, John Walker, and Peter Grabosky, "An Enforcement Taxonomy of Regulatory Agencies," *Law and Policy* 9, no. 3 (July 1987): 325–351.

20. See, for example, May and Winter, "Reconsidering Styles of Regulatory Enforcement"; May and Winter, "Regulatory Enforcement and Compliance"; Peter J. May and Raymond J. Burby, "Making Sense Out of Regulatory Enforcement," *Law and Policy* 20, no. 2 (April 1998): 157–182; Robert A. Kagan, "Regulatory Enforcement," in *Handbook of Regulation and Administrative Law*, ed. David Rosenbloom and Richard D. Schwartz (New York: Marcel Decker, 1994), 383–422; Pautz and Rinfret, *The Lilliputians*; Bridget M. Hutter, "Variations in Regulatory Enforcement Styles," *Law and Policy* 11, no. 2 (April 1989): 153–174; Hutter, *Compliance*.

21. Lo, Liu, and Li, "Dynamics of Inspectors'," 99.

22. See, for example, Keith Hawkins, *Environment and Enforcement: Regulation and the Social Definition of Pollution* (Oxford: Clarendon Press, 1984); Reiss, "Selecting Strategies."

23. See, for example, Pautz and Rinfret, *The Lilliputians*; May and Burby, "Making Sense"; May and Winter, "Reconsidering Styles."

24. Pautz and Rinfret, *The Lilliputians*, 66.

25. Lo, Liu, and Li, "Dynamics of Inspectors'," 100–102.

26. Ibid.

27. May and Winter, "Regulatory Enforcement Styles," 235.

28. See, for example, Pautz and Rinfret, *The Lilliputians*; Pautz, "Trust between Regulators."

29. Robert A. Kagan, Neil Gunningham, and Dorothy Thornton, "Fear, Duty, and Regulatory Compliance: Lessons from Three Research Projects," in *Explaining Compliance: Business Responses to Regulation*, ed. Christine Parker and Vibeke Lehmann Nielsen (Northampton, MA: Edward Elgar, 2011), 37. See also Hutter, *Compliance*; Lauren Edelman and Shauhin A. Talesh, "To Comply or Not to Comply—That Isn't the Question: How Organizations Construct the Meaning of Compliance," in *Explaining Compliance: Business Responses to Regulation*, ed. Christine Parker and Vibeke Lehmann Nielsen (Northampton, MA: Edward Elgar, 2011), 103.

30. Kagan et al., "Fear, Duty, and Regulatory Compliance," 39.

31. See, for example, Pautz and Rinfret, *The Lilliputians*.

32. Peter Mascini and Eelco Van Wijk, 2009, as quoted in Kim Loyens, Carina Schott, and Trui Steen, "Strict Enforcement or Responsive Regulation? How Inspector-Inspectee Interaction and Inspectors' Role Identity Shape Decision Making," in *Inspectors and Enforcement at the Front Line of Government*, ed. Steven Van de Walle and Nadine Raaphorst (New York: Palgrave, 2019), 81.

33. Ian Ayers and John Braithwaite, *Responsive Regulation: Transcending the Deregulation Debate* (New York: Oxford University Press, 1992), 39.

Suggested Readings

Parker, Christine, and Vibeke Lehmann Nielsen, eds. *Explaining Compliance: Business Responses to Regulation*. Northampton, MA: Edward Elgar, 2011.

Pautz, Michelle C., and Sara R. Rinfret. *The Lilliputians of Environmental Regulation: The Perspective of State Regulators*. New York: Routledge, 2013.

Scheberle, Denise. *Industrial Disasters and Environmental Policy: Stories of Villains, Heroes, and the Rest of Us*. New York: Routledge, 2018.

Van de Walle, Steven, and Nadine Raaphorst, eds. *Inspectors and Enforcement at the Front Line of Government*. New York: Palgrave Macmillan, 2019.

Bibliography

Ayers, Ian, and John Braithwaite. *Responsive Regulation: Transcending the Deregulation Debate*. New York: Oxford University Press, 1992.

Bardach, Eugene, and Robert A. Kagan. *Going by the Book: The Problem of Regulatory Unreasonableness*. New Brunswick, NJ: Transaction, 2003.

Braithwaite, John, John Walker, and Peter Grabosky. "An Enforcement Taxonomy of Regulatory Agencies." *Law and Policy* 9 (July 1987): 325–351.

Edelman, Lauren, and Shauhin A. Talesh. "To Comply or Not to Comply—That Isn't the Question: How Organizations Construct the Meaning of Compliance." In *Explaining Compliance: Business Responses to Regulation*, edited by Christine Parker and Vibeke Lehmann Nielsen, 103–122. Northampton, MA: Edward Elgar, 2011.

Fineman, Stephen. "Street-Level Bureaucrats and the Social Construction of Environmental Control." *Organization Studies* 19, no. 6 (1998): 953–974.

Goodsell, Charles T. *The New Case for Bureaucracy*. Thousand Oaks, CA: Sage/ CQ Press, 2014.

Gormley, William T. "Regulatory Enforcement Styles." *Political Research Quarterly* 51, no. 2 (1998): 363–383.

Hall, Jeffrey B., Joakim Lindgren, and Moritz G. Sowada. "Inspectors as Information-Seekers." In *Inspectors and Enforcement at the Front Line of Government*, edited by Steven Van de Walle and Nadine Raaphorst, 35–58. New York: Palgrave, 2019.

Hawkins, Keith. *Environment and Enforcement: Regulation and the Social Definition of Pollution*. Oxford: Clarendon Press, 1984.

Hutter, Bridget M. *Compliance: Regulation and Environment*. Oxford: Clarendon Press, 1997.

———. "Variations in Regulatory Enforcement Styles." *Law and Policy* 11, no. 2 (April 1989): 153–174.

Kagan, Robert A. "Regulatory Enforcement." In *Handbook of Regulation and Administrative Law*, edited by David Rosenbloom and Richard D. Schwartz, 383–422. New York: Marcel Decker, 1994.

Kagan, Robert A., Neil Gunningham, and Dorothy Thornton. "Fear, Duty, and Regulatory Compliance: Lessons from Three Research Projects." In *Explaining Compliance: Business Responses to Regulation*, edited by Christine Parker and Vibeke Lehmann Nielsen, 37–58. Northampton, MA: Edward Elgar, 2011.

Lipsky, Michael. *Street-Level Bureaucracy: Dilemmas of the Individual in Public Services*. New York: Russell Sage Foundation, 1980.

Lo, Carlos, Ning Liu, and Pansy Li. "Dynamics of Inspectors' Enforcement Styles." In *Inspectors and Enforcement at the Front Line of Government*, edited by Steven Van de Walle and Nadine Raaphorst, 95–123. New York: Palgrave, 2019.

Loyens, Kim, Carina Schott, and Trui Steen. "Strict Enforcement or Responsive Regulation? How Inspector-Inspectee Interaction and Inspectors' Role Identity Shape Decision Making." In *Inspectors and Enforcement at the Front Line of Government*, edited by Steven Van de Walle and Nadine Raaphorst, 79–94. New York: Palgrave, 2019.

Mascini, Peter, and Eelco Van Wijk. "Responsive Regulation at the Dutch Food and Consumer Product Safety Authority: An Empirical Assessment of Assumptions Underlying the Theory." *Regulation and Governance* 3, no. 1 (2009): 27–47.

May, Peter J., and Raymond J. Burby. "Making Sense out of Regulatory Enforcement." *Law and Policy* 20, no. 2 (April 1998): 157–182.

May, Peter J., and Søren Winter. "Reconsidering Styles of Regulatory Enforcement: Patterns in Danish Agro-Environmental Inspection." *Law and Policy* 22, no. 2 (April 2000): 143–173.

———. "Regulatory Enforcement and Compliance: Examining Danish Agro-Environmental Policy." *Journal of Policy Analysis and Management* 18, no. 4 (1999): 625–651.

———. "Regulatory Enforcement Styles and Compliance." In *Explaining Compliance: Business Responses to Regulation*, edited by Christine Parker and Vibeke Lehmann Nielsen, 222–245. Northampton, MA: Edward Elgar, 2011.

May, Peter J., and Robert S. Wood. "At the Regulatory Front Lines: Inspectors' Enforcement Styles and Regulatory Compliance." *Journal of Public Administration Research and Theory* 13, no. 2 (2003): 117–139.

Maynard-Moody, Steven, and Michael Musheno. *Cops, Teachers, Counselors: Stories from the Front Lines of Public Service.* Ann Arbor: University of Michigan Press, 2003.

Parker, Christine, and Vibeke Lehmann Nielsen. "Introduction." In *Explaining Compliance: Business Responses to Regulation*, edited by Christine Parker and Vibeke Lehmann Nielsen, 1–33. Northampton, MA: Edward Elgar, 2011.

Pautz, Michelle C. "Front-Line Regulators and Their Approach to Environmental Regulation in Southwest Ohio." *Review of Policy Research* 27, no. 6 (November 2010): 761–780.

———. "Trust between Regulators and the Regulated: A Case Study of Environmental Inspectors and Facility Personnel in Virginia." *Politics and Policy* 37, no. 5 (October 2009): 1047–1072.

Pautz, Michelle C., and Sara R. Rinfret. *The Lilliputians of Environmental Regulation: The Perspective of State Regulators.* New York: Routledge, 2013.

———. "State Environmental Regulators: Perspectives of Trust with Their Regulatory Counterparts." *Journal of Public Affairs* 16, no. 1 (February 2016): 28–38.

Raaphorst, Nadine. "Studying Uncertainty in Decision-Making by Street-Level Inspectors." In *Inspectors and Enforcement at the Front Line of Government*, edited by Steven Van de Walle and Nadine Raaphorst, 11–33. New York: Palgrave, 2019.

Reiss, Albert. "Selecting Strategies of Social Control over Organizational Life." In *Enforcing Regulation*, edited by Keith Hawkins and John Thomas, 23–25. Boston: Kluwer-Nijhoff, 1984.

Riccucci, Norma M. *How Management Matters: Street-Level Bureaucrats and Welfare Reform.* Washington, DC: Georgetown University Press, 2005.

Rutz, Suzanne, and Antoinette de Bont. "Collective Discretionary Room: How Inspectors Decide with Providers and Citizens." In *Inspectors and Enforcement at the Front Line of Government*, edited by Steven Van de Walle and Nadine Raaphorst, 187–204. New York: Palgrave, 2019.

Scholz, John T. "Managing Regulatory Enforcement." In *Handbook of Regulation and Administrative Law*, edited by David Rosenbloom and Richard D. Schwartz, 423–463. New York: Marcel Decker, 1994.

Suquet, Jean-Baptiste. "Are You a True Offender? Bus Ticket Inspection as Deviance Enactment." In *Inspectors and Enforcement at the Front Line of Government*, edited by Steven Van de Walle and Nadine Raaphorst, 59–78. New York: Palgrave, 2019.

Van de Walle, Steven, and Nadine Raaphorst, eds. *Inspectors and Enforcement at the Front Line of Government*. New York: Palgrave Macmillan, 2019.

———. "Introduction: The Social Dynamics of Daily Inspection Work." In *Inspectors and Enforcement at the Front Line of Government*, edited by Steven Van de Walle and Nadine Raaphorst, 1–10. New York: Palgrave Macmillan, 2019.

6

Lessons from the States

Oil and Gas Regulation in Colorado

Robert J. Duffy

ouses are not supposed to explode, but that is exactly what happened on Twilight Avenue in Firestone, Colorado, in the late afternoon of April 17, 2017. The explosion leveled the house, took the lives of Joseph Irwin III and his brother-in law Mark Martinez, and seriously injured Martinez's wife, Erin, and their child. The blast and flames were so intense that the house next door was severely damaged as well and was later demolished.

With a focus on the state of Colorado, this chapter examines the often-overlooked role of state agencies and county and local governments in oil and gas development. Just as federal regulators and inspectors play key roles in shaping environmental policy, so, too, do state actors—especially in areas that federal policy does not address, such as hydraulic fracturing (fracking). Oil and gas development, and the question of how best to regulate it, has arguably been the most controversial issue in Colorado in the twenty-first century. As drilling activity began to ramp up in the late 1990s, much of it in areas unaccustomed to large-scale energy development, the issue moved to the top of the state's political agenda. This chapter examines how the ensuing political conflict has reshaped the state agencies responsible for oil and gas regulation as well as the regulations themselves.

Since 2005, the Colorado Oil and Gas Conservation Commission (COGCC), the state agency primarily responsible for regulating oil and gas development, has been reorganized twice, and the state's permitting rules have been in a state of near-constant flux. A new state law made sweeping changes to the agency, its mission, and its rulemaking powers and granted local and county governments new authority to regulate energy development. This chapter's case study of oil and gas regulation in Colorado illustrates how state, county, and local officials grapple with an unsettled and contentious policy issue and the roles these officials play in shaping and implementing environmental policy.

Case Study: Oil and Gas Regulation in Colorado

The investigation into the Firestone explosion revealed the probable cause was the ignition of fugitive natural gas from a cut flow line attached to an active oil and gas well about 170 feet from the home. Located about seven feet underground, the flow lines run between the wellhead and collection tanks, which had been removed years earlier.[1] Investigators concluded that a mix of propane and methane seeped into the house through French drains and a sump pump, and it ignited when Irwin and Martinez were replacing a water heater in the basement. A report by the National Transportation Safety Board concluded that the line, which was located only six feet from the home's foundation, had likely been accidently severed during construction two years earlier.[2] The well was originally opened in 1993 but had been out of service for more than a year when it was restarted at the end of January 2017.[3] Gas was apparently leaking into the soil from that point on, but since it had not been treated, it had no odor and was therefore undetectable.

The investigation also revealed the gas lines had not been handled according to state regulations, which required that lines be disconnected from the source, cleared of all traces of liquid fuels, and capped at both ends.[4] In fact, the National Transportation Safety Board report noted that none of the lines near the residence had been properly abandoned. The agency also concluded that the decision by local authorities to allow houses to be built on land adjacent to or that had previously been part of an oil and gas production field without complete documentation from the operator on the location and status of its pipelines contributed to the accident.[5]

The incident also revealed deficiencies in the monitoring and inspection of wells and pipelines—problems that had been highlighted previously in a report issued by state regulators in 2014. The report had been mandated by the state legislature, which wanted a risk-based assessment of the state's oil and gas inspections program. According to the report, although corroding pipelines were the source of at least half of the failures leading to oil and gas spills, the state did not have a "formal program to monitor ongoing compliance" regarding integrity testing for pipelines in the state.[6] The *Denver Post* later reported that it took the state nearly a year to institute such a program, when the COGCC hired three employees to ensure the safety of thousands of miles of pipeline associated with the state's fifty-three thousand active wells and more than thirty-six thousand inactive wells. According to the *Post*, by the end of 2017, the agency's new inspector had conducted approximately four hundred inspections, and the new state pipeline engineer had audited just twenty-four of the largest oil and gas operators to verify that they kept records documenting whether they had conducted required annual pipeline pressure tests. State regulators, in short, had very little insight into what energy companies, especially smaller firms, were doing to maintain the integrity of their pipelines.[7]

In the weeks and months following the accident, the governor, state legislature, local and county officials, environmental organizations, the industry, and the state agency responsible for regulating oil and gas development grappled with how to respond. Ten days after the explosion, Anandarko Petroleum, the operator of the well, announced it would temporarily shut down more than three thousand similar wells in Weld County.[8] Later that month, Governor John Hickenlooper ordered that all flow lines within one thousand feet of occupied buildings be inspected within thirty days and be tested for integrity within sixty days.

The Firestone accident prompted regulators to alter the rules governing flow lines. In February 2018, the COGCC issued new rules governing how flow lines would be established, tested, and abandoned. The new rules required operators to test smaller flow lines that had previously been exempted from testing and to file forms with the COGCC documenting flow line locations. Under the new rules, local governments, but not the general public, would have access to specific location data. Access to such data had been hotly contested, with those pushing for a more significant local role arguing that the public had a right to know where gas lines might be. The industry countered that making such information widely available posed a potential safety risk.[9]

More broadly, the explosion and subsequent investigation renewed long-standing public concerns about the safety of oil and gas development in the state, especially in densely populated areas. Since the oil and gas boom that began in the mid-1990s, energy issues have been a prominent issue for the public and for policy makers. Oil and gas bills have been a staple in every session of the state legislature, and multiple governors have been forced to address the issue. Several local communities enacted moratoriums or bans on fracking, which the state then challenged in court. In November 2018, voters had the chance to weigh in, as they defeated a statewide ballot measure that would have imposed larger setbacks for oil and gas activity. The oil and gas industry spent heavily to defeat the measure, but their victory was short-lived; in spring 2019, the state legislature enacted a sweeping overhaul of oil and gas regulation, and since then state and county officials have been working toward the adoption of new rules. How that rulemaking plays out will have enormous implications for oil and gas regulation in the state.

An Overview of Oil and Gas Regulation in Colorado

The laws governing energy exploration and development in Colorado, many of which date back fifty to one hundred years, are similar to those in other western states in that they seek to encourage and promote energy exploration and development. Colorado law, for example, recognizes separate ownership of the surface estate and the mineral estate and the distinct private property rights associated with each. Generally speaking, mineral rights take priority over surface rights. In the case of oil and gas resources, the presumption in Colorado is that oil and gas companies that have purchased or leased mineral rights are entitled to exercise their property rights and develop the resource. Accordingly, state law provides for access to the mineral estate by allowing subsurface owners "reasonable use" of the surface estate.

Similarly, for more than a century, Colorado courts have recognized the rule of accommodation, which acknowledges that the right to extract gas is limited by a duty to minimize impacts for the surface owner. In *Gerrity Oil and Gas Corporation v. Magness* (1997), the state supreme court upheld the precedent, but it has had little practical effect because no state agency actively enforces it. Finally, although Colorado does have a law similar to the National Environmental Policy Act, which requires agencies to perform environmental impact statements, until very recently agencies were not required to perform "cumulative impact" analyses for projects on state or private lands. As a result, state agencies review projects on a well-by-well basis, which understates their overall impact on the environment and public health.

Oil and gas wells on private and state lands in Colorado require permits from three state agencies. The first is the Water Quality Control Division of the Colorado Department of Public Health and Environment, which regulates the discharge of pollutants into the state's surface water and groundwater and administers the state's drinking water regulations. Second, the Colorado state engineer issues permits for the diversion of groundwater. Third, the Colorado Oil and Gas Conservation Commission regulates wells for spacing, density, construction, and safety. Because most of the current controversy over energy in the state revolves around the COGCC, the remainder of this chapter focuses on that commission's activities.

Earlier chapters have explained that Congress delegates considerable authority to federal regulatory agencies. Delegation of authority also occurs in the states, as state legislatures grant policy-making power to state regulatory agencies. State legislatures delegate authority for the same reasons as Congress. First, state legislators do not want to specify programmatic details in legislation because doing so could be detrimental to their reelection efforts. Second, members may craft ambiguous laws so that administrative agencies can use their expertise to interpret and implement the law. Presumably, state agencies are composed of experts in specific policy areas and have the knowledge to make sound policy decisions. In short, state legislatures create state-level policy in the same way the Congress creates federal policy, and state rulemaking is of considerable importance in the absence of federal policy. For instance, the lack of federal policy for fracking means it is up to states to use their rulemaking processes to establish their own policies.

Federal and state rulemaking procedures are very similar and are designed to encourage public participation in governmental decision making, in the hopes that greater participation will lead to not only better policy but also decisions that are more legitimate in the eyes of the public. Rulemaking procedures also provide standards for judicial review if a party believes she or he has been harmed by an agency action.

As discussed in Chapter 1, every executive branch agency in Colorado must follow the rulemaking procedures outlined in the Administrative Procedure Act unless explicitly exempted.[10] As at the federal level, the state rulemaking process involves four stages: issuance of a notice of proposed rulemaking (NPRM), a comment period on the proposed rule, a hearing on the proposed rule, and final adoption of the rule.

When an agency wants to issue a new rule, or amend an existing one, it must first file an NPRM with the secretary of state. The secretary of state then publishes the notice in the *Colorado Register*, which serves as notice to the public. By law, agencies are also required to notify any individuals who have informed the agency that they want to be notified of any proposed rulemaking. Once the NPRM is filed with the secretary of state, the public may offer comments on the proposed rule. In Colorado, state agencies must accept and consider comments from the public before that agency can adopt, amend, or otherwise change any regulation not explicitly exempted from the Administrative Procedure Act. The third stage involves hearings on the proposed rule, where the agency accepts testimony (oral or written) from supporters or critics. Following the hearing, the agency has 180 days to file adopted rules with the secretary of state for publication in the *Colorado Register*; rules take effect twenty days after publication or on such later date as is stated in the rule.[11]

Throughout much of its history, the Colorado Oil and Gas Conservation Commission worked to facilitate energy exploration and development in the state. In large part, this was a direct result of the agency's statutory mandate to foster the development, production, and use of oil and gas resources. Although the commission's mandate was revised in 2007 to include a responsibility to conduct its work in a "manner consistent with the protection of public health, safety, and welfare, including protection of the environment and wildlife resources," critics alleged that it did not alter the agency's clear sense of mission and that it consistently emphasized development at the expense of its regulatory duties.[12] In explaining its actions, the commission said that its authority "to prevent and mitigate significant environmental harm did not override its obligation to encourage development of oil and gas resources."[13]

Beginning in the mid-1990s, this stance put the commission at odds not only with surface owners but also with many local communities, which sought to use their authority to regulate land use within their own boundaries by imposing some restrictions on energy development. In response to charges that it had failed to require companies to spend money on equipment that might reduce any environmental impacts, in 2007 the commission noted that the law that requires it to protect public health, safety, and welfare also requires it to take "into consideration cost-effectiveness and technical feasibility."[14] Faced with a burgeoning revolt by local and county governments, the COGCC in 2001 sought to claim absolute control of the permitting process by issuing a new rule that stated, in part, "The permit to drill shall be binding with respect to any conflicting local government permit or land use approval process."[15] In defending the rule, which was unsuccessfully challenged in court, the agency's then director, Rich Griebling, stated, "The Legislature has articulated (that) there is some value in having oil and gas regulated consistently statewide, and not to have 40-some counties with differing regulations."[16]

In their efforts to reform energy policy in Colorado, drilling opponents tried a variety of institutional strategies, including numerous attempts to revise the COGCC's rules and rulemaking procedures as well as attacks on its jurisdictional monopoly. Initially, this entailed expanding the scope of the conflict by attracting the attention and support of allies in local and county governments, the courts, and ultimately the state legislature. In 1985, for example, the city of Greeley, which had been growing rapidly, adopted an ordinance banning oil and gas wells inside the city limits. The industry challenged the move in court and won. The court ruled the city could not ban all drilling and said that the state had an overriding interest to make sure that energy resources were not "wasted" because they were not being developed. Over the years, the state's courts have ruled that both the oil and gas commission and local governments have jurisdiction over energy development, but they have struggled to offer a definitive stance as to the precise division of authority.

Critics did have some success in the courts, though. In December 2006, a state appeals court unanimously ruled that counties could regulate oil and gas operations if their regulations did not conflict with state policies. The case involved a dispute between an energy firm and Gunnison County, which had issued regulations requiring firms to obtain a permit before beginning operations. The appeals court said that if local rules could be harmonized with state laws, the counties could regulate oil and gas development. However, the court also noted that the rules would not stand if the court found the rules would "materially impede or destroy the state interest."[17]

In addition to their efforts to chip away at the COGCC's jurisdiction via the courts and county commissions, reformers tried to move the conflict up the ladder of authority to the state legislature. Beginning in the late 1990s, reformers and their allies introduced numerous bills to rewrite the statutory framework governing energy development in the state. As in other states, most of these efforts failed. Among the most common were measures that tried to increase the bonding and reclamation requirements for well operators, measures that would alter the composition of the COGCC, and others that would have required energy producers to obtain a surface damage agreement from individual surface owners before commencing operations. Not surprisingly, the COGCC and the oil and gas lobby strenuously resisted such changes, and most were either substantially weakened during debate or were easily defeated in one or both chambers.

The breakthrough came in 2007, and it can be traced primarily to the change in partisan control of state government. In 2004, Democrats won control of the state legislature—something that had not happened in nearly four decades. In the 2006 elections, Democrats gained additional seats and also won control of the governorship. With both branches now in Democratic hands, both sides had to adjust their long-term political calculations and strategies.

The most significant legislative change involved the composition of the COGCC. Of specific concern was the statutory requirement that five of the seven members of the COGCC had once worked for the oil and gas industry. After years of inaction, in 2007 the legislature approved a compromise measure enlarging the commission from seven to nine members while reducing the number that had industry ties from five to three. In addition, the law broadened the commission's makeup by adding the executive directors of the state's natural resources and public health departments. The measure also mandated that other members should represent a range of interests, including local government, environmental or wildlife protection, agricultural production, and royalty owners.[18] Although the energy industry objected to the measure, it dropped its opposition after a provision enhancing the commission's authority to deny access to mineral rights was eliminated.[19]

As noted above, the same law also revised the legislative declaration in the Oil and Gas Conservation Act, which governs the COGCC. The original statute noted that it was in the public interest to "foster, encourage, and promote the development, production, and utilization of oil and gas". The amended version deleted the words *encourage* and *promote* and instead declared it to be in the public interest to "foster the *responsible, balanced* development, production, and utilization" of the state's oil and gas resources "in a manner consistent with protection of public health, safety, and welfare, including *protection of the environment and wildlife resources*" (emphasis in original). It also required the agency to submit semiannual reports to the state's general assembly tracking the number of applications for permits to drill, the average time of review, and a description of the number and character of applications for permits for which approval was withheld. Finally, the general assembly reserved for itself the power to review, alter or repeal any COGCC rules promulgated pursuant to the law.[20]

The reformulation of the COGCC did little to stem the conflict over oil and gas in the state. If anything, the conflicts increased in intensity. Over the next decade, several communities, frustrated with what they perceived as the state's failure to protect public health and safety, even adopted moratoriums or outright bans on oil and gas development. To cite one example, in 2012, the Longmont city council passed an ordinance banning oil and gas development within residential neighborhoods and imposed several restrictions stricter than those of the state. The city said its actions were consistent with its land-use powers, but both the industry and the COGCC sued, claiming the city ordinances were preempted by state authority. It was the first time the state sued a city over oil and gas regulation.[21] The Longmont ordinance, along with efforts in Fort Collins, Thornton, and other communities to restrict oil and gas drilling, were struck down by the state's supreme court, citing the state's clear primacy over these matters.

Some steps were taken in the next few years toward more stringent state regulation. A new rule established disclosure standards for the fluid pumped into wells in fracking operations, and another rule increased the setbacks of drill rigs from 150 feet to 500 feet. In 2014, pursuant to an executive order by former governor Bill Ritter, the state began tracking greenhouse gas emissions. That same year, the COGCC issued rules requiring drillers to find and repair methane leaks. It is unclear, however, whether those rules actually reduced emissions, because the agency does not consistently monitor the effects of those repairs. In addition, the state lacks reliable methods for tracking methane emissions, with some studies indicating that emissions were declining and others finding increases.[22]

In 2014, Governor Hickenlooper created a blue-ribbon task force to study the issue of local control and prepare recommendations for the state legislature to adopt. The task force met over five months, toured sites, held public hearings throughout the state, and considered fifty-six proposals. The group's focus was on resolving conflicts between local and state regulation of oil and gas and how to handle drilling within communities. In the end, the group issued nine recommendations aimed at reducing conflict between local communities and energy development. The nine final recommendations included establishing processes for greater communication and collaboration between industry and local governments, more inspections of drill sites, increased staffing for the COGCC and Department of Public Health and Environment, and additional research into possible health effects of drilling.[23]

If anyone believed the task force would settle the oil and gas wars, they were sorely mistaken. Industry representatives thought the proposal gave local communities too much say, while those who supported tougher oil and gas rules were disappointed because the task force's final recommendations did not do enough to ensure public health and safety and to enhance local control over drilling.[24] Democrats in the state legislature introduced at least seven bills related to oil and gas in the next sessions, but all were blocked by the Republican senate. The conflict reached new heights when, after threatening to do so for years, environmental and citizen groups gathered more than 120,000 signatures to place on the November 2018 ballot a measure that sought to impose tougher restrictions on drilling. Proposition 112, which was ultimately defeated 57–43 by voters, would have increased the setback distance between wells and homes from five hundred feet to twenty-five hundred feet. The measure would also have increased the setbacks for wells located near drinking water sources, playgrounds, and parks. The industry spent more than $40 million opposing the measure, arguing that it would effectively "wipe out the industry" by barring drilling on most of the state's land. The industry also contended that the measure would destroy thousands of jobs and deprive the state of millions of dollars in tax revenue.[25]

At the same time, the courts continued to be dragged into the dispute over energy development. For the most part, state courts consistently sided with the COGCC (and the industry) and its claims of state primacy in permitting and other regulatory decisions. But in 2017, the Colorado Court of Appeals upset the apple cart, overturning a trial-court decision and siding with industry critics who argued that the COGCC could issue a permit only if it had first concluded that a new well would not harm the environment. That decision set the stage for a much-anticipated ruling by the state supreme court in early January 2019, in what came to be known as the *Martinez* case.

In a unanimous decision, the court ruled that the appellate court had erred in interpreting the law governing the COGCC's duties. Under that law, the COGCC's role was to "foster the responsible, balanced development, production, and utilization of the natural resources of oil and gas in the state of Colorado in a manner consistent with protection of public health, safety, and welfare, including protection of the environment and wildlife resources." The appellate court stated that the words *consistent with* meant the agency had to decide that development would protect public health, safety and welfare before approving new permits. The state supreme court disagreed, stating that the COGCC must seek only to balance competing interests, noting the language of the law "envisions some possible environmental and public health risks," and that the agency should seek to mitigate environmental harms while balancing "cost-effectiveness and technical feasibility."[26]

Although Proposition 112 was defeated in November 2018, it did attract a significant level of public support and showed that the issue was still potent politically. In that same election, Democrats maintained control of the state house of representatives, took control of the state senate, and swept all major statewide races. Among the victors was Jared Polis, a longtime critic of drilling who supported local governments having more authority in energy regulation; Polis won the governorship. Suddenly, the political terrain had shifted, and the industry could no longer rely on the Republican senate to protect its interests. In a sign of what was to come, house speaker K. C. Becker stressed the importance of passing new oil and gas legislation in her opening speech to the chamber in 2019.[27]

Senate Bill 19-181: The Protect Public Welfare Oil and Gas Operations Act

After months of contentious debate and intense interest group lobbying, in April 2019 the Colorado General Assembly adopted Senate Bill 19-181 (SB 181), the Protect Public Welfare Oil and Gas Operations Act. The law, adopted on largely party-line votes, fundamentally altered oil and gas regulation in the state. Among other things, SB 181 changed the primary mission of the COGCC from fostering the development of oil and gas resources to prioritizing the protection of public health, safety, welfare, the environment, and wildlife. The law also reorganized the COGCC and required the agency to conduct several significant rulemakings by July 2020. Finally, the new law effectively ended state primacy and granted local governments the authority to regulate oil and gas development through their land use and planning powers. Importantly, SB 181 gave local governments the power to impose rules that exceed state requirements, but it did not give them the power to be less restrictive than the state.

Perhaps the most significant change mandated by SB 181 was shifting the agency's core mission from fostering energy development to protecting public health, safety, welfare, and the environment. The change was critical because the statutory mission to foster development had so long defined the agency's approach to its regulatory responsibilities. Critics had argued that the charge was a fatal conflict of interest—how could an agency responsible for promoting an industry also be counted on to regulate it? The second significant change under the new law was the weakening of state primacy and granting local governments considerable new authority to regulate a wide range of oil and gas activities as part of their traditional land use powers. SB 181 requires that operators file an application to drill with the local government before filing for a state permit and to disclose the local government's decision in the state application. The law also specifically allows local governments to regulate the location and impacts of drilling, such as noise and odors. Finally, SB 181 allows local governments to inspect oil and natural gas facilities, impose fines for violations, and impose fees on drillers to cover regulatory costs.[28]

The law further specifies that local governments may enact regulatory standards that are more stringent than the state's, but they are not allowed to make the standards weaker. To ensure against this outcome, the legislation requires operators to obtain permits from both the local government and the COGCC; the COGCC has the authority to deny a permit if it believes the local government has not done enough to protect public health and safety.[29]

As noted above, SB 181 altered the composition of the COGCC for the second time in fifteen years, further weakening industry's role and broadening representation of other interests. After some wrangling, the legislature declared that the agency would now consist of seven full-time members, appointed by the governor with the consent of the state senate. The shift from a part-time to full-time body was made, in part, at the behest of the industry, which preferred that decisions be made by a professional panel with some expertise in oil and gas matters. The directors of the state departments of natural resources and public health and environment would serve as nonvoting members. The other members were to include one representative each from the oil and gas industry, the environmental and public health communities, and royalty owners. The law also called for geographic representation on the commission to include locations that rely on oil and gas revenue as well as those affected by energy development.[30]

The new law delegates considerable discretion to the COGCC and requires the agency to promulgate several significant new rules by July 2020—an ambitious goal. The agency was directed to review and amend its flow line rules, increase public disclosure of well location and information, and specify when inactive infrastructure must be reinspected before being put back into use.[31] The law also called for the COGCC to amend its rules to move from fostering development to regulating in a manner that is protective of public health, safety, welfare, environment, and wildlife.[32] The law also directs the agency to adopt rules that require developers to consider alternative locations for proposed oil and natural gas facilities that are near populated areas.[33] A fourth mandated rulemaking, to be conducted with the Colorado Department of Public Health and Environment, is to evaluate and address the potential cumulative impacts of oil and gas development.[34] A fifth rulemaking would aim to ensure wellbore integrity and more rigorous testing to require that all wells meet current standards.[35]

As one might expect, the industry vigorously opposed the bill, even going so far as to bus in workers from oil and gas fields as a show of force during early hearings on the bill. The industry also paid for radio and TV ads slamming the bill, saying it would "shut down energy production" in the state and that it could damage Colorado's economy for years to come.[36] Republican lawmakers tried to stop the bill, but their minority status meant they could only delay it—although they were able to win some concessions in the final version.

In a signing ceremony for the bill, Governor Polis said hopefully, "Today, with the signing of this bill, it is our hope that the oil and gas wars that have enveloped our state are over." A spokesperson for Colorado Rising said the bill represented "the most substantial shift we have seen in decades and puts communities on much better footing when confronted with industrial oil and gas in their backyards." A spokesperson for the Colorado Petroleum Council, on the other hand, said the bill was "deeply flawed" but expressed thanks that the final bill was not as bad as it could have been because lawmakers worked to address some of their concerns.[37]

Oil and Gas Regulation after SB 181

The passage of SB 181 meant that several state agencies had to draft new rules to comply with the law, and they had to do so quickly. The nine-member Colorado Air Quality Control Commission, which oversees the state's air quality programs, was charged with helping the COGCC protect public health and safety and with issuing new requirements for emissions tracking. The emissions data would then be incorporated into the state's greenhouse gas inventory. In December 2019, the air quality control commission approved tougher, uniform statewide air quality standards. The rules require twice yearly inspections for wells within one thousand feet of schools, homes, and other public facilities; emissions reductions from storage tanks; and annual emissions reports.[38] The commission's air pollution control division also proposed new rules requiring oil and gas operators to monitor and repair methane leaks more frequently; to obtain permits during the first ninety days of drilling; to monitor emissions of methane and volatile organic compounds, which contribute to ground-level ozone; and to report all emissions to the state.[39]

In response to public complaints about headaches, dizziness, and other health issues, the department of public health and environment commissioned a report examining the public health risks facing residents living near oil and gas wells.[40] The preliminary findings suggested that people living within two thousand feet of wells faced a higher short-term risk from exposure to benzene and other chemicals during preproduction stages. In response, the COGCC announced that it would immediately begin monitoring emissions near wells and would begin subjecting new wells within two thousand feet of homes to greater regulatory scrutiny. The agency also said that it would encourage companies that had recently been authorized to drill within two thousand feet of homes to minimize emissions.[41]

No agency was more affected by the new law than the COGCC. Under the law, the COGCC was required to approve most of its new rules by July 1, 2020. After that, a full-time commission was to take over for the volunteer members. As a result, the COGCC spent much of 2019–2020 holding public hearings and conducting the key rulemakings mandated by the new law.[42]

One month after Governor Polis signed the bill into law, the COGCC issued a list of sixteen "objective criteria" it would employ to determine whether applications to drill would warrant additional review. The criteria included any applications for oil and gas wells within a municipality; locations within fifteen hundred feet of a high-occupancy building, municipal or county boundary, or plotted subdivision; locations within two thousand feet of a school; and locations within floodplains, public drinking water supplies, or water resource areas. Applications to drill in areas where a relevant local government requests additional information are also subject to additional review.[43] At the time, there were more than six thousand pending drilling permits at the COGCC; industry officials expressed concern that the additional reviews would mean lengthy delays in permit approvals. Indeed, in an interview, the agency's director said that some permits might have to wait up to one year until the agency completes its mandated rulemakings under SB 181.[44]

At the same time, the COGCC had to confront another controversial issue: whether it should halt the issuance of new drilling permits until the rulemaking had been completed and the agency had decided what criteria would be used to determine which drilling permits would be subject to a higher level of scrutiny under the law. Drilling opponents argued that the agency should delay all its permitting work, but that position was rejected by both Dan Gibbs, head of the state department of natural resources, and COGCC director Jeff Robbins. Both men said the law did not support a de facto moratorium on permitting.[45]

The agency's first rulemaking involved proposed changes to its administrative procedures, known as Series 500. The process was quite contentious, with ninety-nine participants weighing in. The proceedings involved motions and countermotions by both industry and drilling opponents, each seeking to disqualify the other from participating. Many of the environmental and community groups argued that the agency's rules of procedures should reflect the change in the COGCC's mission from fostering oil and gas development to the protection of the environment, public health, and wildlife. Robbins disagreed, saying that doing so would be "outside the scope" of the rulemaking.[46]

In December 2019, the COGCC completed its first major substantive rulemaking implementing SB 181. It was far less contentious. The law specifically charged the agency with conducting a rulemaking that would "allow the public disclosure of flowline information" and to determine when deactivated flow lines would need to be inspected. The resulting flow line rule was informed by months of public hearings and a public comment process involving fifty-five parties. In the words of COGCC director Jeff Robbins, "We worked with environmental, industry, local government, homeowners and other stakeholders from across the state in a collaborative manner to arrive at sensible solutions that will deliver stronger protections and more accurate and publicly available mapping information, and help to increase public safety as a result of this flowline rulemaking."[47]

The new rules allow the state to map thousands of miles of underground flow lines and to make public the general location of such lines. The rule also imposes new requirements for testing the integrity of flow lines and ensuring that lines are safely shut down. After Firestone, critics had pushed for greater transparency in the location of oil and gas lines, but the industry expressed concerns about security, vandalism, and landowner privacy. The rule seeks to satisfy both sides by requiring public disclosure of the lines within a range of twenty-five feet; the state will have more precise information about location, but that information will be accessible to the public only by checking with either the COGCC or local governments. The agency considered requiring operators to remove abandoned lines once they had been disconnected from a well and capped, but it ultimately decided that abandoned lines could remain in place after operators consulted with the COGCC director and if an independent expert verified that all procedures were followed.[48]

The agency's second substantive rulemaking—on well integrity— was also far less contentious than one might think. Industry trade groups negotiated the draft rules with environmental organizations, with assistance from COGCC staff. The draft rules, which were ratified by the COGCC in June 2020, were the result of discussions between the Colorado Oil and Gas Association, the Colorado Petroleum Council, the Environmental Defense Fund, the Sierra Club, and Earthjustice. The rules required stronger protections for wells up to three thousand feet below the surface that are drilled through aquifers used for drinking water. Under the new rules, each of the state's roughly fifty-two thousand wells would be pressure-tested annually to ensure that the wells were intact and working properly. The rules would also impose more stringent requirements for well casings and cement; more specifically, the rules would require additional testing of steel casing and cement work and would limit the use of chemicals in the drilling muds used in areas where groundwater was to be protected.

Although many of the environmental groups had sought to require the same protections for deeper wells, the Environmental Defense Fund expressed hopes that the new rule would be a model for other states. Similarly, industry representatives expressed support for both the process and the rule's substance, although this support was not unanimous. Smaller oil and gas operators expressed concerns that the testing requirements under the new rules would impose disproportionate burdens on them relative to larger firms. In developing the rules, COGCC staff worked with staff from the water quality control division to ensure that rule changes were consistent with the Colorado Water Quality Control Act and the COGCC's duties under the new law. COGCC representatives also expressed satisfaction with the collaborative rulemaking.[49]

Several other significant rules remain on the COGCC's docket, including rulemakings to require companies to demonstrate that they have sufficient funds to safely abandon wells and others to address the cumulative impacts of energy development.[50] Although much remains to be done, thus far, it seems the COGCC is aiming to solicit input from all key stakeholders and to promulgate rules that balance competing claims. It remains to be seen how successful that effort will be.

County and Local Actions after SB 181

After decades of a near state monopoly, the new law gives county and local governments more authority over energy development in their communities. What seems clear is that communities will use that authority very differently; thus far, some municipalities have opted to impose more stringent controls on the industry, while others have sought to encourage further development. The increased responsibility creates challenges for local officials, who will have to find the resources and expertise to develop the new rules and to carry out whatever inspections, monitoring, and enforcement that ensue. The shifting regulatory landscape also creates challenges for the industry, which now must navigate the various rules that communities will adopt rather than one set of statewide rules. In short, all the major actors in this issue will have to muddle through as the new roles, rules, and responsibilities are sorted out.

Indeed, there was flurry of county and local government activity in the months after SB 181 became law. By mid-August 2019, the communities of Berthoud, Broomfield, Erie, Lafayette, Superior, and Timnath, as well as Boulder County, had either imposed or extended moratoriums on drilling.[51] Officials in other counties and municipalities embraced the new authority granted by SB 181 to develop new, more restrictive rules governing oil and gas activity. Some hired inspectors to conduct site visits. Still others, such as the city of Aurora and Weld County, made it clear that they would use their authority to encourage new drilling activity.

Commissioners in Adams County, the site of intense conflicts between new home building and energy development, adopted a six-month moratorium on new oil and gas wells just after the bill was introduced. According to the commission's chair, the county was not seeking to ban energy development but was "intent on protecting health and safety first and allowing as many minerals as to be extracted as possible."[52] After holding public hearings through the summer, the county eventually issued new rules in September, ending the moratorium. The new rules covered well location and siting, air and water quality, noise and dust standards, reclamation procedures, and bonding requirements, and imposed a one-thousand-foot setback from buildings, twice the distance required by state rules. Industry groups immediately denounced the new regulations, saying they were unreasonable and exceeded what was allowed under the new law.[53]

In Weld County, on the other hand, county commissioners have taken a markedly different approach to the new law, and to oil and gas development more broadly. Weld County is situated above the Denver-Julesburg Basin, the largest oil and gas field in the state, and produces nearly 90 percent of the oil and more than one-third of the natural gas in the state.[54] As a result, Weld County is heavily dependent on oil and gas revenue—much more so than most counties. In fact, more than half of the county's property tax revenue comes from the oil and gas industry. Not surprisingly, elected officials from the county, mostly Republicans, were adamantly opposed to SB 181, claiming it would lead to a decline in tax revenue and dramatic service reductions and that it could bankrupt the state.[55]

Weld County commissioners had acted even before SB 181 to facilitate energy development. In November 2016, the commission voted unanimously to adopt new, less restrictive oil and gas rules, eliminating the public hearing requirements for new projects in residential areas. In addition, companies would no longer be required to obtain permission from landowners affected by projects but would only have to show that they had tried to do so.[56]

After the passage of SB 181, the commission seemed to reverse course and embrace the idea of local control. In July 2019, it voted unanimously to make unincorporated areas of the county an "area of state interest"—a step designed to allow the county to exert control over surface regulations for oil and gas drilling permits. At the same time, the commissioners created a new oil and gas energy department, which it claimed would "firmly establish the county's local control over mineral resources in unincorporated Weld County," and in the process, circumvent state control.[57] In defending its actions, the commission cited a 1974 law that allowed local governments "to identify, designate, and regulate areas and activities of state interest through a local permitting process. The general intention of these powers is to allow for local governments to maintain their control over particular development projects even where the development project has statewide impacts."[58] The commission also cited provisions of SB 181 that said, "Each local government within its respective jurisdiction has the authority to plan for and regulate the use of land by . . . regulating the surface impacts of oil and gas operations." Under this interpretation, Weld County could regulate setbacks, noise control, and other contentious issues, while the COGCC would regulate underground issues, such as pipeline depth and integrity.[59] In short, the county argued that its rules would take precedence over the state's if there were differences.

In response to these actions, the COGCC sent a letter to the commission noting that it would continue to regulate oil and gas activity in the unincorporated parts of the county despite the creation of a new county department. According to the letter, "While SB 19-181 provides local governments with siting authority over oil and gas surface locations, it does not diminish the COGCC's authority to regulate the orderly development of oil and gas throughout the state." The letter also noted, "Under SB 19-181, local governments may impose regulations that are 'more protective than state requirements,' but they are not authorized to bypass the COGCC's regulations." The county commissioners responded with a news release the same day arguing that the COGCC letter "contradicts SB 19-181 and, as we read it, attempts to take the 'local' out of 'local control. . . . For the COGCC, through the State Attorney's Office, to now not only attempt to assume local control from local governments but also require a duplicative process for the oil and gas industry is both disappointing and irresponsible."[60]

If this exchange proves anything, it is that the conflict over oil and gas regulation in Colorado is not over.

Concluding Thoughts

Previous chapters have focused on the role of federal actors in environmental policy making. This chapter, as well as Chapter 5, demonstrates that state actors play a central role in shaping and implementing environmental policy. As oil and gas development has increased in Colorado, so has the political conflict surrounding it. As a result, state policy has been in flux since 2007, as governors, the state legislature, state courts, and a host of local and county governments have struggled to sort out an acceptable division of labor. The state rulemaking environment is complicated by multiple actors, different policy goals, and shifting agency missions and regulatory standards.

For regulators at the state and local levels, it has proven to be a difficult task. Over time, the tide has shifted in favor of those seeking greater oversight of the industry, which claimed that the new law would make the state less attractive for developers and would slow permitting, eliminate jobs, and dramatically reduce tax revenue flowing to government coffers. Because rulemaking is ongoing, it will take some time to sort out the effects of the new rules and to know what effect it has on energy activity in the state. Whatever happens, state regulators will play a central role in shaping oil and gas policy. Chapter 7 examines how state and federal regulators have worked to manage the problem of coal ash, a significant problem in some regions of the United States.

Notes

1. Flow lines typically carry oil or gas from individual wellheads to gathering or pumping stations; in fields with multiple wells, separate flow lines converge at a central facility. Primary responsibility for regulating and inspecting flow lines belongs to the COGCC. Gathering lines are the next stage of the process—they carry oil and gas from gathering facilities to processing stations or to market. They are regulated by the Colorado Public Utility Commission. Interstate transmission lines, which are larger and carry oil and gas even longer distances, are regulated by the U.S. Department of Transportation, which monitors the safety of operating pipelines, and the Federal Energy Regulatory Commission, which handles approvals, permitting, and siting for new pipe-

lines. See Jacquelyn Pless, "Making State Gas Pipelines Safe and Reliable: An Assessment of State Policy," National Conference of State Legislatures, March 2011, https://www.ncsl.org/research/energy/state-gas-pipelines-federal-and-state-responsibili.aspx.

2. Kevin Vaughan, "Uncapped, Abandoned Gas Line Caused Firestone Home Explosion," KUSA (Denver) *9News*, May 5, 2017, https://www.9news.com/article/news/investigations/uncapped-abandoned-gas-line-caused-firestone-home-explosion/73-436094693; Christopher N. Osher and Bruce Finley, "Oil and Gas Industry Pipeline Problems Are Well-Established. Why Did It Take a Fatal Explosion to Spur Action?" *Denver Post*, December 18, 2017, https://www.denverpost/2017/05/07/firestone-explosion-raises-questions-pipeline-risks/; National Transportation Safety Board, "Executive Summary: Pipeline Accident Brief, Natural Gas Explosion at Family Residence," January 2019, https://ntsb.gov/investigations/AccidentReports/Pages/PAB1902.aspx.

3. See Vaughan, "Uncapped, Abandoned Gas Line."

4. Grace Hood, "Two Years after a House Exploded Near a Firestone Oil and Gas Well, the Federal Report Is Finally Here," *Colorado Public Radio*, October 29, 2019, https://www.cpr.org/2019/10/29/ntsb-firestone-house-explosion-report/.

5. See National Transportation Safety Board, "Executive Summary"; Hood, "Two Years."

6. National Transportation Safety Board.

7. Osher and Finley, "Oil and Gas Industry."

8. See Vaughan, "Uncapped, Abandoned Gas Line"; "Timeline of Fatal Gas Explosion at House in Firestone," *CBS4 News Denver*, June 4, 2017, https://denver.cbslocal.com/guide/timeline-firestone-explosion/.

9. Grace Hood, "What's Changed since the Firestone Explosion? Let's Count the New Regulations," *Colorado Public Radio*, April 16, 2018, https://www.cpr.org/2018/04/16/whats-changed-since-the-firestone-explosion-lets-count-the-new-regulations/#.WtS4nnj_lyw.twitter.

10. State Administrative Procedure Act, C.R.S. § 24-4-101.

11. Colorado Office of Policy, Research, and Regulatory Reform, "What to Know about Colorado's Rulemaking and Cost-Benefit Analysis Process," 2019, https://www.colorado.gov/pacific/dora-oprrr/coprrr-process.

12. Oil and Gas Conservation Act, C.R.S. § 34-60-101, et seq. (2007).

13. Colorado Oil and Gas Conservation Commission, "Typical Questions from the Public about Oil and Gas Development in Colorado," 2007, http://www.oil-gas.state.co.us/.

14. Ibid.

15. Ray Ring, "Backlash: Local Governments Tackle an In-Your-Face Rush to Coalbed Methane," *High Country News*, September 2, 2002, https://www.hcn.org/issues/233/11371.

16. Ibid.

17. Howard Pankratz, "Counties Win Oil, Gas Ruling," *Denver Post*, December 14, 2006, https://www.denverpost.com/2006/12/14/counties-win-oil-gas-ruling/.

18. Oil and Gas Conservation Act, C.R.S. § 36-60-104 (2)(a) (2007).

19. Jeri Clausing, "Oil-Commission Revamp Gains: All Sides on Board," *Denver Post*, April 17, 2007, http://www.denverpost.com/search/ci_5682183; Jeri Clausing, "Senate Backs Changes to State's Oil and Gas Commission," *Denver Post*, April 25, 2007, http://www.denverpost.com/search/ci_5743223.

20. Oil and Gas Conservation Act, C.R.S. § 36-60-106 (2007).

21. Scott Rochat and Longmont Times-Call, "State Sues Longmont over Oil and Gas Drilling Rules," *Longmont Times-Call*, July 30, 2007, https://www.timescall.com/2012/07/30/state-sues-longmont-over-oil-and-gas-drilling-regulations/.

22. Mark Jaffe and John Frank, "What the Landmark Oil and Gas Bill Really Says—And Its Significance for Colorado," *Colorado Sun*, March 5, 2019, https://coloradosun.com/2019/03/05/whats-in-colorado-oil-and-gas-reform-legislation/; John Herrick, "Colorado Will Consider Requiring Drillers to Track and Report Methane Emissions," *Colorado Independent*, July 9, 2019, https://www.coloradoindependent.com/2019/07/09/colorado-oil-gas-methane-monitoring-reporting/.

23. Keystone Center, "Colorado Oil and Gas Task Force Final Report," February 27, 2015, https://firebasestorage.googleapis.com/v0/b/torid-heat-3070.appspot.com/o/Programs%2FOGTF%2FOilGasTaskForceFinalReport.pdf?alt=media&token=6f8a500d-bd36-4c46-a8c7-fca21971771c.

24. Cathy Proctor, "Colorado Oil and Gas Task Force Recommendations Disappointing for Many," *Denver Business Journal*, February 25, 2015, https://www.bizjournals.com/denver/blog/earth_to_power/2015/02/colorado-oil-and-gas-task-force-recommendations.html.

25. Grace Hood, "The Ballot Fight to Push Back Oil and Gas Is All about Winning Hearts and Minds," *Colorado Public Radio*, October 12, 2018, https://www.cpr.org/2018/10/12/the-ballot-fight-to-push-back-oil-and-gas-is-all-about-winning-hearts-and-minds/. See also Mark Jaffe, "Proposition 112 Fails, but Big Vote Total Signals Oil and Gas Setbacks Will Be Headed to the Colorado Capitol," *Colorado Sun*, November 7, 2018, https://coloradosun.com/2018/11/07/prop-112-fails-oil-gas-fight-statehouse/.

26. *COGCC v. Martinez*; John Ingold and John Frank, "Colorado Supreme Court Sides with Oil and Gas in Major Ruling on Environmental Regulations," *Colorado Sun*, January 14, 2019, https://coloradosun.com/2019/01/14/martinez-supreme-court-decision/; John Ingold, Mark Jaffe, and John Frank, "Three Ways of Looking at the Colorado Supreme Court's Major Oil and Gas Ruling," *Colorado Sun*, January 15, 2019, https://coloradosun.com/2019/01/15/colorado-supreme-court-martinez-oil-gas-explained/.

27. Mark Jaffe, "Oil Rigs May Draw Protests, but Colorado's Pipelines Could Get Pinched by Democratic Lawmakers," *Colorado Sun*, January 10, 2019, https://coloradosun.com/2019/01/10/colorado-oil-and-gas-midstream-2019-legislature/.

28. Jesse Paul, "Gov. Polis Signs Democrats' Sweeping Oil and Gas Bill into Law, Marking Major Shift in Regulatory Authority over Drilling," *Colorado Sun*, April 16, 2019, https://coloradosun.com/2019/04/16/senate-bill-181-oil-gas-law

-colorado-signed/; John Frank, "The Oil and Gas Bill Is Nearing Final Approval: Here's a Look at the Concessions the Industry Won," *Colorado Sun*, April 1, 2019, https://coloradosun.com/2019/04/01/oil-gas-legislation-industry-changes/.

29. John Aguilar, "Adams County Tightens Oil and Gas Rules, First to Do So since Senate Bill 181 Passed," *Denver Post*, September 3, 2019, https://www.denverpost.com/2019/09/03/oil-gas-adams-county-colorado/.

30. See Frank, "The Oil and Gas Bill"; see also Grace Hood, "After Colorado Lawmakers Usher in Sweeping Oil and Gas Rules, It'll Be Up to COGCC's Jeff Robbins to Implement Them," *Colorado Public Radio*, April 11, 2019, https://www.cpr.org/show-segment/after-colorado-lawmakers-usher-in-sweeping-oil-and-gas-rules-itll-be-up-to-cogccs-jeff-robbins-to-implement-them/.

31. Oil and Gas Conservation Act, C.R.S. § 34-60-106 (2019).

32. Oil and Gas Conservation Act, C.R.S. § 34-60-106(2.5)(a) (2019).

33. U.S. Energy Information Administration, "*Today in Energy: Colorado Changes Its Regulatory Structure for Oil and Natural Gas Production*," June 27, 2019, https://www.eia.gov/todayinenergy/detail.php?id=39993; Jaffe and Frank, "What the Landmark."

34. Oil and Gas Conservation Act, C.R.S. § 34-60-106(11)(c)(II) (2019).

35. Oil and Gas Conservation Act, C.R.S. § 34-60-106(18) (2019).

36. American Petroleum Institute, "Stop S.B. 181 in Colorado," https://www.youtube.com/watch?v=DVHu8nxKTHE.

37. Paul, "Governor Polis." See also Mark Jaffe, "New Energy Legislation Didn't End Colorado's 'Oil and Gas Wars.' It Just Relocated Them," *Colorado Sun*, June 14, 2019, https://coloradosun.com/2019/06/14/colorado-oil-gas-wars-rulemaking/.

38. Associated Press, "Colorado Regulators Approve Tougher Oil and Gas Air Rules, Ending Statewide Disparity," *Colorado Sun*, December 20, 2019, https://coloradosun.com/2019/12/20/oil-gas-regulations-changes-colorado/.

39. John Herrick, "Colorado Will Consider Requiring Drillers to Track and Report Methane Emissions," *Colorado Independent*, July 9, 2019, https://www.coloradoindependent.com/2019/07/09/colorado-oil-gas-methane-monitoring-reporting/; John Herrick, "Oil and Gas Emissions 'Not Acceptable,' Colorado's Top Air Quality Regulator Says," *Colorado Independent*, July 29, 2019, https://www.coloradoindependent.com/2019/07/29/oil-gas-emissions-regulations/.

40. ICF, "Final Report: Human Health Risk Assessment for Oil and Gas Operations in Colorado," October 17, 2019, https://drive.google.com/file/d/1pO41DJMXw9sD1NjR_OKyBJP5NCb-AO0I/view.

41. Bruce Finley, "Colorado to Tighten Oversight of Oil and Gas Sites Near Homes in Wake of Study Finding Possible Short-Term Health Effects," *Denver Post*, October 17, 2019, https://www.denverpost.com/2019/10/17/colorado-oil-gas-health-risks-study/; Judith Kohler, "New State Rules Will Create First-Ever Public Mapping of Underground Oil and Gas Lines," *Denver Post*, November 21, 2019, https://www.denverpost.com/2019/11/21/colorado-adopts-rule-mapping-oil-gas-lines/.

42. Judith Kohler, "Six Months after Colorado's Sweeping Oil and Gas Law Took Effect, Fight over Path Forward Hasn't Faded," *Longmont Times-Call*, October 24, 2019, https://www.timescall.com/2019/10/24/colorado-oil-gas-law-6-months-old-sb-181/.

43. Colorado Oil and Gas Conservation Commission, "SB 19-181 Required Director Objective Criteria," May 16, 2019, http://cogcc.state.co.us/documents/sb19181/DOC/COGCC_Directors_Final_Objective_Criteria_20190516.pdf.

44. John Herrick, "Jeff Robbins, State's Top Oil and Gas Regulator, Talks Local Control and Setbacks," *Colorado Independent*, April 29, 2019, https://www.coloradoindependent.com/2019/04/29/colorado-jeff-robbins-cogcc-oil-gas-regulations/.

45. Dan Elliott, "Oil and Gas Activists Want Colorado to Pause New Drilling While Regulations Are Being Sorted Out," *Colorado Sun*, May 15, 2019, https://coloradosun.com/2019/05/15/new-oil-and-gas-laws-rollout-colorado/. See also Dan Elliott, "Colorado Regulators Won't Halt Oil and Gas Drilling as They Implement New Law Prioritizing Safety," *Colorado Sun*, May 21, 2019, https://coloradosun.com/2019/05/21/colorado-new-oil-and-gas-regulations-drilling-permits/.

46. Elliot, "Colorado Regulators."

47. Cynthia Wilson, "Colorado Oil and Gas Conservation Commission Completes First Rulemaking under SB 19-181," *North Forty News*, December 11, 2019, https://northfortynews.com/colorado-oil-gas-conservation-commission-completes-first-rulemaking-under-sb-19-181/.

48. Kohler, "New State Rules."

49. Mark Jaffe, "Colorado Drillers, Enviros Near Accord on Groundwater Protection Rules. The Only Rub: How Much to Protect?" *Colorado Sun*, February 6, 2020, https://coloradosun.com/2020/02/06/colorado-oil-gas-environment-groundwater-rules/.

50. Kohler, "Six Months."

51. Chase Woodruff, "How Colorado Towns Are Starting to Use Their 'Local Control' of Fracking," *Westword*, July 5, 2019, https://www.westword.com/news/how-colorado-towns-are-starting-to-use-local-control-of-fracking-11399130/.

52. Mark Jaffe, "Colorado Oil and Gas Overhaul Tips Power toward Local Government. How That Power Is Used Will Vary Widely," *Colorado Sun*, April 9, 2019, https://coloradosun.com/2019/04/09/colorado-oil-gas-local-control/.

53. Aguilar, "Adams County Tightens"; see also "Adams County Approves New Oil and Gas Regulations after 6-month Moratorium on New Development," KDVR, September 3, 2019, https://kdvr.com/2019/09/03/adams-county-approves-new-oil-and-gas-regulations-after-6-month-moratorium-on-permits/.

54. Weld County Colorado, "Oil and Gas Energy Department," 2020, https://www.weldgov.com/departments/oil_and_gas_energy.

55. Sherrie Peif, "Weld County Asserts Local Control over Oil and Gas Regulations; Say They Are Holding Governor to His Promise," *Complete Colorado*, June 10, 2019, https://pagetwo.completecolorado.com/2019/06/10/weld-county-asserts

-local-control-over-oil-and-gas-regulations-say-they-are-holding-governor-to
-his-promise/.

56. Catherine Sweeney, "Weld County Officials Adopt New Oil and Gas Regulations," *Greeley Tribune*, November 28, 2016, https://www.greeleytribune .com/news/local/weld-county-officials-adopt-new-oil-and-gas-regulations/.

57. Weld County Colorado, "Oil and Gas Energy Department."

58. See Peif, "Weld County Asserts."

59. Ibid.

60. Trevor Reid, "Weld Officials: COGCC Letter Illustrates Desire to Suppress Local Control of Oil and Gas," *Greeley Tribune*, July 22, 2019, https://www .greeleytribune.com/news/cogcc-state-will-still-regulate-oil-and-gas-in-weld -county-regardless-of-new-local-department/.

Suggested Readings

"Colorado Oil and Gas Rule Seeks to Limit Federal Uncertainty." *Inside EPA's Water Policy Report* 14, no. 9 (2005): 9–10.

Davis, Charles E. "The Politics of 'Fracking': Regulating Natural Gas Drilling Practices in Colorado and Texas." *Review of Policy Research* 29, no. 2 (2012): 177–191.

Feriancek, Jean. "Local Regulation of Mineral Extraction in Colorado." *Natural Resources and Environment* 24, no. 4 (2010): 51–52. Available at www.jstor .org/stable/25802040.

Kansal, Tushar, and Patrick Field. "States' Diverse Approaches to Allowing Local Regulation of Oil and Gas Development." In *Approaches to Local Regulation of Shale Gas Development*, 20–35. Working paper. Cambridge, MA: Lincoln Institute of Land Policy, 2013. Available at www.jstor.org/stable/resrep18441.6.

Malin, Stephanie A., Stacia Ryder, and Peter M. Hall. 2018. "Contested Colorado: Shifting Regulations and Public Responses to Unconventional Oil Production in the Niobrara Shale Region." In *Fractured Communities: Risk, Impacts, and Protest against Hydraulic Fracking in U.S. Shale Regions*, edited by Anthony E. Ladd, 198–223. New Brunswick, NJ: Rutgers University Press, 2018.

Bibliography

"Adams County Approves New Oil and Gas Regulations after 6-Month Moratorium on New Development." KDVR. September 3, 2019. Available at https://kdvr .com/2019/09/03/adams-county-approves-new-oil-and-gas-regulations-after-6 -month-moratorium-on-permits/.

Aguilar, John. "Adams County Tightens Oil and Gas Rules, First to Do So since Senate Bill 181 Passed." *Denver Post*, September 3, 2019. Available at https:// www.denverpost.com/2019/09/03/oil-gas-adams-county-colorado/.

Associated Press. "Colorado Regulators Approve Tougher Oil and Gas Air Rules, Ending Statewide Disparity." *Colorado Sun*, December 20, 2019. Available at https://coloradosun.com/2019/12/20/oil-gas-regulations-changes-colorado/.

Clausing, Jeri. "Oil-Commission Revamp Gains: All Sides on Board." *Denver Post*, April 17, 2007. Available at http://www.denverpost.com/search/ci_5682183.

———. "Senate Backs Changes to State's Oil and Gas Commission." *Denver Post*, April 25, 2007. Available at http://www.denverpost.com/search/ci_5743223.

Colorado Office of Policy, Research and Regulatory Reform. Colorado's Rule-making and Cost-Benefit Analysis Process. 2020. Available at https://www.colorado.gov/pacific/dora-oprrr/coprrr-process.

Colorado Oil and Gas Conservation Commission. SB 19-181 Required Director Objective Criteria. May 16, 2019. Available at http://cogcc.state.co.us/documents/sb19181/DOC/COGCC_Directors_Final_Objective_Criteria_20190516.pdf.

———. "Typical Questions from the Public about Oil and Gas Development in Colorado." 2007. Available at http://www.oil-gas.state.co.us/.

Colorado Supreme Court. No. 17SC297, COGCC v. Martinez. 2019. Available at https://www.courts.state.co.us/userfiles/file/Court_Probation/Supreme_Court/Opinions/2017/17SC297.pdf.

Elliott, Dan. "Colorado Regulators Won't Halt Oil and Gas Drilling as They Implement New Law Prioritizing Safety." *Colorado Sun*, May 21, 2019. Available at https://coloradosun.com/2019/05/21/colorado-new-oil-and-gas-regulations-drilling-permits/.

———. "Oil and Gas Activists Want Colorado to Pause New Drilling while Regulations Are Being Sorted Out. *Colorado Sun*, May 15, 2019. Available at https://coloradosun.com/2019/05/15/new-oil-and-gas-laws-rollout-colorado/.

Finley, Bruce. "Colorado to Tighten Oversight of Oil and Gas Sites Near Homes in Wake of Study Finding Possible Short-Term Health Effects." *Denver Post*, October 17, 2019. Available at https://www.denverpost.com/2019/10/17/colorado-oil-gas-health-risks-study/.

Frank, John. "The Oil and Gas Bill Is Nearing Final Approval. Here's a Look at the Concessions the Industry Won." *Colorado Sun*, April 1, 2019. Available at https://coloradosun.com/2019/04/01/oil-gas-legislation-industry-changes/.

Herrick, John. "Colorado Will Consider Requiring Drillers to Track and Report Methane Emissions." *Colorado Independent*, July 9, 2019. Available at https://www.coloradoindependent.com/2019/07/09/colorado-oil-gas-methane-monitoring-reporting/.

———. "Jeff Robbins, State's Top Oil and Gas Regulator, Talks Local Control and Setbacks." *Colorado Independent*, April 29, 2019. Available at https://www.coloradoindependent.com/2019/04/29/colorado-jeff-robbins-cogcc-oil-gas-regulations/.

———. "Oil and Gas Emissions 'Not Acceptable,' Colorado's Top Air Quality Regulator Says." *Colorado Independent*, July 29, 2019. Available at https://www.coloradoindependent.com/2019/07/29/oil-gas-emissions-regulations/.

Hood, Grace. "After Colorado Lawmakers Usher In Sweeping Oil and Gas Rules, It'll Be Up to COGCC's Jeff Robbins to Implement Them." *Colorado Public Radio*, April 11, 2019. Available at https://www.cpr.org/show-segment/after-colorado-lawmakers-usher-in-sweeping-oil-and-gas-rules-itll-be-up-to-cogccs-jeff-robbins-to-implement-them/.

———. "The Ballot Fight to Push Back Oil and Gas Is All about Winning Hearts and Minds." *Colorado Public Radio*, October 12, 2018. Available at https://www.cpr.org/2018/10/12/the-ballot-fight-to-push-back-oil-and-gas-is-all-about-winning-hearts-and-minds/.

———. "2 Years after a House Exploded near a Firestone Oil and Gas Well, the Federal Report Is Finally Here." *Colorado Public Radio*, October 29, 2019. Available at https://www.cpr.org/2019/10/29/ntsb-firestone-house-explosion-report/.

———. "What's Changed since the Firestone Explosion? Let's Count the New Regulations." *Colorado Public Radio*, April 16, 2018. Available at https://www.cpr.org/2018/04/16/whats-changed-since-the-firestone-explosion-lets-count-the-new-regulations/#.WtS4nnj_lyw.twitter.

ICF. "Final Report: Human Health Risk Assessment for Oil and Gas Operations in Colorado." October 17, 2019. Available at https://drive.google.com/file/d/1pO41DJMXw9sD1NjR_OKyBJP5NCb-AO0I/view.

Ingold, John, and John Frank. "Colorado Supreme Court Sides with Oil and Gas in Major Ruling on Environmental Regulations." *Colorado Sun*, January 14, 2019. Available at https://coloradosun.com/2019/01/14/martinez-supreme-court-decision/.

Ingold, John, Mark Jaffe, and John Frank. "Three Ways of Looking at the Colorado Supreme Court's Major Oil and Gas Ruling." *Colorado Sun*, January 15, 2019. Available at https://coloradosun.com/2019/01/15/colorado-supreme-court-martinez-oil-gas-explained/.

Jaffe, Mark. "Colorado Drillers, Enviros Near Accord on Groundwater Protection Rules. The Only Rub: How Much to Protect?" *Colorado Sun*, February 6, 2020. Available at https://coloradosun.com/2020/02/06/colorado-oil-gas-environment-groundwater-rules/.

———. "Colorado Oil and Gas Overhaul Tips Power toward Local Government. How That Power Is Used Will Vary Widely." *Colorado Sun*, April 9, 2019. Available at https://coloradosun.com/2019/04/09/colorado-oil-gas-local-control/.

———. "New Energy Legislation Didn't End Colorado's 'Oil and Gas Wars.' It Just Relocated Them." *Colorado Sun*, June 14, 2019. Available at https://coloradosun.com/2019/06/14/colorado-oil-gas-wars-rulemaking/.

———. "Oil Rigs May Draw Protests, but Colorado's Pipelines Could Get Pinched by Democratic Lawmakers." *Colorado Sun*, January 10, 2019. Available at https://coloradosun.com/2019/01/10/colorado-oil-and-gas-midstream-2019-legislature/.

———. "Proposition 112 Fails, but Big Vote Total Signals Oil and Gas Setbacks Will Be Headed to the Colorado Capitol." *Colorado Sun*, November 7, 2018.

Available at https://coloradosun.com/2018/11/07/prop-112-fails-oil-gas-fight
-statehouse/.

Jaffe, Mark, and John Frank. "What the Landmark Oil and Gas Bill Really Says—
And Its Significance for Colorado." *Colorado Sun*, March 5, 2019. Available at
https://coloradosun.com/2019/03/05/whats-in-colorado-oil-and-gas-reform
-legislation/.

Keystone Center. "Colorado Oil and Gas Task Force Final Report." February
27, 2015. Available at https://firebasestorage.googleapis.com/v0/b/torid-heat
-3070.appspot.com/o/Programs%2FOGTF%2FOilGasTaskForceFinalReport
.pdf?alt=media&token=6f8a500d-bd36-4c46-a8c7-fca21971771c.

Kohler, Judith. "New State Rules Will Create First-Ever Public Mapping of
Underground Oil Gas Lines." *Denver Post*, November 21, 2019. Available at
https://www.denverpost.com/2019/11/21/colorado-adopts-rule-mapping-oil
-gas-lines/.

———. "Six Months after Colorado's Sweeping Oil and Gas Law Took Effect, Fight
over Path Forward Hasn't Faded." *Times-Call*, October 24, 2019. Available at
https://www.timescall.com/2019/10/24/colorado-oil-gas-law-6-months-old
-sb-181/.

National Transportation Safety Board. "Executive Summary: Pipeline Accident
Brief: Natural Gas Explosion at Family Residence." January 2019. Available
at https://ntsb.gov/investigations/AccidentReports/Pages/PAB1902.aspx.

Osher, Christopher N., and Bruce Finley. "Oil and Gas Industry Pipeline Problems
Are Well-Established. Why Did It Take a Fatal Explosion to Spur Action?"
Denver Post, December 18, 2017. Available at https://www.denverpost.com
/2017/05/07/firestone-explosion-raises-questions-pipeline-risks/.

Pankratz, Howard. "Counties Win Oil, Gas Ruling." *Denver Post*, December 14,
2006. Available at http://www.denverpost.com/search/ci_4842396.

Paul, Jesse. "Gov. Polis Signs Democrats' Sweeping Oil and Gas Bill into Law,
Marking Major Shift in Regulatory Authority over Drilling." *Colorado Sun*,
April 16, 2019. Available at https://coloradosun.com/2019/04/16/senate-bill
-181-oil-gas-law-colorado-signed/.

Peif, Sherrie. "Weld County Asserts Local Control over Oil and Gas Regulations;
Say They Are Holding Governor to His Promise." *Complete Colorado*, June
10, 2019. Available at https://pagetwo.completecolorado.com/2019/06/10
/weld-county-asserts-local-control-over-oil-and-gas-regulations-say-they
-are-holding-governor-to-his-promise/.

Pless, Jacquelyn. "Making State Gas Pipelines Safe and Reliable: An Assessment
of State Policy." National Conference of State Legislatures. March 2011.
Available at https://www.ncsl.org/research/energy/state-gas-pipelines-fed
eral-and-state-responsibili.aspx.

Proctor, Cathy. "Colorado Oil and Gas Task Force Recommendations Disappointing for Many." *Denver Business Journal*, February 25, 2015. Available at https://
www.bizjournals.com/denver/blog/earth_to_power/2015/02/colorado-oil-and
-gas-task-force-recommendations.html.

Reid, Trevor. "Weld Officials: COGCC Letter Illustrates Desire to Suppress Local Control of Oil and Gas." *Greeley Tribune*, July 22, 2019. Available at https://www.greeleytribune.com/news/cogcc-state-will-still-regulate-oil-and-gas-in-weld-county-regardless-of-new-local-department/.

Ring, Ray. "Backlash: Local Governments Tackle an In-Your-Face Rush to Coalbed Methane." *High Country News*, September 2, 2002.

Rochat, Scott, and Longmont Times-Call. "State Sues Longmont over Oil and Gas Drilling Rules." *Longmont Times-Call*, July 30, 2012. Available at https://www.timescall.com/2012/07/30/state-sues-longmont-over-oil-and-gas-drilling-regulations/.

Sweeney, Catherine. "Weld County Officials Adopt New Oil and Gas Regulations." *Greeley Tribune*, November 28, 2016. Available at https://www.greeleytribune.com/news/local/weld-county-officials-adopt-new-oil-and-gas-regulations/.

"Timeline of Fatal Gas Explosion at House in Firestone." *CBS4 News Denver*, June 4, 2017. Available at https://denver.cbslocal.com/guide/timeline-firestone-explosion/.

U.S. Energy Information Administration. "Today in Energy: Colorado Changes Its Regulatory Structure for Oil and Natural Gas Production." June 27, 2019. Available at https://www.eia.gov/todayinenergy/detail.php?id=39993.

Vaughan, Kevin. "Uncapped, Abandoned Gas Line Caused Firestone Home Explosion." KUSA (Denver) *9News*, May 5, 2017. Available at https://www.9news.com/article/news/investigations/uncapped-abandoned-gas-line-caused-firestone-home-explosion/73-436094693.

Weld County Colorado. Oil and Gas Energy Department. *2020. Available at* https://www.weldgov.com/departments/oil_and_gas_energy.

Wilson, Cynthia. "Colorado Oil and Gas Conservation Commission Completes First Rulemaking under SB 19-181." *North Forty News*, December 11, 2019. Available at https://northfortynews.com/colorado-oil-gas-conservation-commission-completes-first-rulemaking-under-sb-19-181/.

Woodruff, Chase. "How Colorado Towns Are Starting to Use Their 'Local Control' of Fracking." *Westword*, July 5, 2019. Available at https://www.westword.com/news/how-colorado-towns-are-starting-to-use-local-control-of-fracking-11399130/.

7

Interacting Authorities

How Southeastern States and the EPA Regulate Coal Ash Disposal

CHARLES DAVIS

Coal as a long-standing fuel source for the generation of electricity has been in a steady decline since 2010 because of policy makers' concerns about its negative impacts on environmental quality and climate change.[1] These concerns are reinforced by the coal industry's inability to effectively compete in the power production marketplace with cleaner fuels such as natural gas and renewable energy resources.[2] Although the extraction or combustion of coal leads to a variety of pollution problems, this chapter focuses on an increasingly important policy issue involving an unwanted by-product of its use—the disposal and storage of coal ash and the risks it poses for water quality and public health.

The emergence of coal ash disposal and containment as a regulatory problem did not occur at the federal level until 2015, when the Environmental Protection Agency (EPA) enacted a coal ash rule during President Obama's second term. At this point, the federal government became involved in a crowded policy space that has been largely dominated by state officials. Key policy players include state regulators, environmental groups, utilities, local officials, and both state and federal courts. Not surprisingly, there is considerable variation in statewide efforts to address storage and disposal issues associated with coal ash.[3] Adding to the complexity of the policy-making process is legal uncertainty that surrounds the relationship between coal ash leaks and the subsequent contamination of underground water sources. Recent federal court rulings suggest the Clean Water Act does not cover pollution that migrates through groundwater, because states maintain jurisdiction over these sources.[4]

Documented throughout this book is the myriad of actors in the implementation of environmental policy, from those involved in the federal rulemaking process to the environmental inspectors charged with implementation at the state level. This chapter provides a better understanding of the regulatory implementation of an environmental policy (water quality) by reducing or mitigating the negative impacts of energy production (i.e., coal). It includes an overview of various state and federal efforts to regulate coal ash from 2008 to the present. One key question considers the extent to which states give priority to environmental or public health goals in relation to financial concerns raised by coal or utility interests. Another consideration is the shift in the intergovernmental context. A greater decision-making role for the EPA since 2015 has left state regulators with less discretionary authority. Just as state officials need to adjust to a new president with different policy priorities, it is equally necessary for state regulators to be flexible in dealing with a change in the governor's office, especially when the governor does not share the same partisan orientation as the president.

As evidenced throughout this book, many factors influence how state regulatory decisions are made. This chapter evaluates the magnitude of pollution problems linked to coal ash spills or leaks (including catastrophic events), dependence on coal as a source of energy and jobs, partisan factors affecting how governors and legislators weigh competing policy values, initiatives undertaken by industry or environmental interests, the changing views of some utility officials, and increasing public awareness of and concern about health risks associated with exposure to coal ash contaminants. As noted in Chapters 5 and 6, the EPA does not enforce environmental requirements, but it does offer guidance on technical and safety standards. However, the agency's importance as a regulator has gradually increased in recent years. Following a discussion of the EPA as a regulatory participant, attention is directed to two states that have been actively involved in coal storage and disposal policies—Kentucky and North Carolina.

Coal Ash as a Regulatory Problem

When coal is burned in power plants to generate electricity, it creates by-products often referred to as coal combustion residuals (CCRs), or coal ash. Coal ash contains fly ash, a powdery form of silica; bottom ash, a coarse substance that remains at the bottom of the coal furnace; and scrubber sludge. According to a survey conducted by the American Coal Ash Association, coal ash is one of the largest sources of industrial waste produced in the United States, totaling about 130 million tons in 2014.[5] Some major utilities such as Duke Energy Corporation highlight coal ash's beneficial use in building materials, such as wallboard, and as an additive in making concrete and cement.[6]

However, serious environmental concerns exist with the storage and disposal of coal ash. CCRs include toxic ingredients such as arsenic, mercury, lead, selenium, and cadmium, which can adversely affect public and ecological health when they are exposed to soil, surface water, or groundwater supplies. For example, lead and mercury are classified as neurotoxins and are known to be associated with developmental problems among at-risk populations, notably children.[7] Other ecological impacts have not received much attention. A study by A. Dennis Lemly and Joseph Skorupa analyzed twenty-one coal waste sites and concluded the price tag for addressing pollution-related damages to fish and wildlife exceeded $2.3 billion.[8]

The problem is severe. A report prepared by the Environmental Integrity Project concluded that unsafe levels of toxic chemicals were found to be affecting groundwater quality at 91 percent of ash disposal sites in the United States.[9] The geographical reach of coal ash as a serious water pollution control issue covers twenty-two states, largely in the Midwest and Southeast, representing roughly three-fourths of the coal-powered utilities in the nation.[10]

Key management concerns include whether coal ash is stored in ponds located close to power plants or in on-site surface impoundments. If these ponds do not have a liner, there is greater risk of coal ash migration to and contamination of underground water sources. A related concern is whether the banks of slurry ponds are unstable. A collapse of a retaining wall designed to hold the slurry in place can easily lead to a major spill that adversely affects both public safety and environmental quality.

However, removing and transferring ash from a pond to an alternative site is expensive. Approaches favored by utility interests include dry storage, which typically means depositing fly ash into landfills or putting a cap or cover on coal slurry deposits. Although these practices are legal in many jurisdictions, they do not necessarily reduce the risks of contamination; neither method requires the installation of a liner or barrier to prevent contaminants from leaching into groundwater. According to Mayor Lee McCarty of Wilsonville, Alabama, the cap-in-place approach is particularly troublesome, because it does nothing more than maintain "pollution in perpetuity."[11]

The EPA Addresses Coal Dust Issues

An early test of the EPA's involvement with coal dust policy issues occurred during congressional debate over the Resource Conservation and Recovery Act (RCRA) of 1976. An important question was whether coal dust should be considered a solid waste along with nonhazardous forms of municipal garbage under the Solid Waste Disposal Act or whether CCRs should be regulated as hazardous waste under Subtitle C of the RCRA. Ultimately, an amendment to the Solid Waste Disposal Act was put forward by Representative Thomas Bevill (D-AL). The amendment as adopted exempted coal dust from regulatory coverage, largely on the grounds that a more restrictive classification under the RCRA would interfere with industry efforts to recycle CCRs for commercially valuable purposes.[12] The Bevill Amendment did require that EPA researchers continue to evaluate risks posed by CCRs to groundwater quality that might result in a reconsideration of federal regulatory concerns.

The transformation of CCRs from a solid waste issue to a serious water pollution concern was bolstered greatly by disastrous spills in 2008 and in 2014. The 2008 spill at the Tennessee Valley Authority's (TVA) coal plant in Kingston, Tennessee, happened after the collapse of a holding pond retaining wall, resulting in the release of over a billion gallons of slurry into the Tennessee River. Drinking water quality for the town was adversely affected, and nearby tributaries and rivers were polluted along with buildings and infrastructure.[13]

Emergency response activities were initiated by the TVA. A command center was established, and the coal plant operator worked with assorted Tennessee state agencies, EPA Region 4 officials, the Federal Emergency Management Agency, and the Army Corps of Engineers to coordinate cleanup activities and assistance for impacted communities. The financial impact of the spill was immense—the price tag for cleanup tasks and damages incurred was more than $1 billion. In 2015, a federal judge approved a $27.8 million settlement from the TVA to compensate affected residents and property owners.[14]

The Kingston incident led President Obama's EPA to develop a CCR assessment program in 2009 to evaluate the structural integrity of dams and surface impoundments associated with coal power plants. In response to an EPA request for information, electric utility officials identified 427 units or ponds containing coal slurry. Of particular concern were units the EPA classified as having a "high hazard" rating—that is, impoundments likely to result in the loss of human life should a breach occur. Forty-four such impoundments in twenty-six different places were subsequently targeted for state inspections to contribute not only to ongoing safety programs but to local emergency preparedness efforts. Utility officials also directed attention to dams with "significant hazard potential" that could result in significant economic damages as well as negative consequences for environmental quality and ecological health.[15]

To address these problems, the EPA proposed a rule in 2010 to manage coal dust storage and disposal standards under the Resource Conservation and Recovery Act. This proposal led to extended commentary over whether the agency should seek a stricter standard under Subtitle C's regulatory process requiring that coal dust be classified as a hazardous waste or whether it should be identified as a solid waste more akin to municipal garbage. The latter interpretation would lead to lower compliance costs and was favored by utility firms, state officials, and the American Coal Ash Association. They touted its beneficial uses that could be more easily realized through recycling than through more expensive disposal options. On the other hand, environmental groups such as Earthjustice and the Southern Environmental Law Center favored a tougher and ecologically sensitive standard.[16]

In 2015, the new rule was issued by EPA administrator Gina McCarthy. Although McCarthy ultimately opted for a balanced approach that initially tilted in the direction of industry interests, the EPA did

choose to impose new structural standards for landfills and impoundments that coal plants would have to meet, along with requirements
of liners for new ponds to protect against leaks. While the new rule
required companies to close unlined slurry ponds and to halt new
shipments of coal ash to deficient facilities, the agency did allow companies more time to make the necessary changes.[17]

Coal plant operators were also required to monitor groundwater
quality and to post improvements on a public website. However, one
regulatory gap of concern to environmentalists was the decision to
exempt older inactive coal ash dumps, thus allowing the possibility
of continuing groundwater contamination from these sources.[18] Another concern was the absence of any enforcement authority for the
EPA under the new regulation. If utilities chose not to comply with
the new standards, it would be up to citizens or state officials to force
compliance through lawsuits.[19]

This led to an important question for members of Congress and
the Obama administration. Should enforcement of coal dust standards be contingent on citizen lawsuits—an approach favored by environmentalists—or on actions taken by state officials—an approach
preferred by industry officials and state regulators? In December
2016, Congress chose the latter approach by enacting the Water Infrastructure Improvements for the Nation (WIIN) Act. The new policy
amended the earlier Resource Conservation and Recovery Act to provide for state coal ash permitting programs. It also gave the EPA the
authority to review and accept state permit programs—a process that
was designed to mirror existing regulations for municipal landfill
programs under Subtitle D of the RCRA.[20]

While the Obama administration placed considerable emphasis
on developing energy policies that were compatible with environmental protection, a marked shift in policy priorities occurred following the election of President Donald J. Trump in 2016. He campaigned as a populist, promising to reverse Obama's "war on coal"
and to restore a central role for fossil fuel use in the nation's energy
policies.[21] Trump then filled his energy and environmental cabinet
positions with pro-industry leaders with little or no public sector experience and began an unprecedented rollback of energy and environmental regulations, ostensibly to spur economic growth.[22]

In 2017, EPA administrators followed up on the WIIN program with the release of a document to provide guidance to states wishing to assume permitting responsibilities. Although this information was well received by industry officials, state regulators, and the Utility Solid Waste Activities Group (a trade association representing utilities), the move was met with skepticism by environmentalists. A major concern was whether states possessed the necessary organizational and financial capacity to implement regulatory tasks related to coal disposal and storage activities. According to Frank Holleman, an attorney with the Southern Environmental Law Center, "State environmental agencies are underfunded, understaffed, and under political influence."[23] To date, the EPA has authorized two states—Oklahoma and Georgia—to operate their own programs. For other states and Native American reservations, the EPA will oversee permit decisions for surface impoundments and landfills.[24]

The EPA also took a number of actions designed to lessen the financial impact of the 2015 coal dust rule on coal plants and utilities. A draft rule issued in July 2018 focused on changes to coal ash management decisions by allowing coal operators to accept a higher volume of waste at slurry ponds and ash pits and giving companies more time to comply with regulatory requirements. The new proposal also allowed for a temporary suspension of groundwater monitoring requirements if state officials decided there was little risk of contaminating nearby aquifers.[25]

These actions prompted litigation from environmental groups concerned about the risks of regulatory backsliding to groundwater quality. On August 21, 2018, the Washington, DC, Circuit Court ruled the agency's actions allowing shipments of coal waste to unlined slurry ponds were illegal and that such impoundments should have been closed (*Utility Solid Wastes Group v. EPA*). In addition, the court held that the regulatory exemption given to inactive ash pits was contrary to the statutory requirements spelled out in the RCRA and should also be altered.[26] Shortly thereafter, a lawsuit initiated by environmentalists contended the above changes should also be applicable to Oklahoma's coal ash management program since the state had recently received permitting authority from the EPA. This argument was affirmed in March 2019 by the DC Circuit Court in *Waterkeepers Alliance v. EPA*.

In November 2019, the EPA issued another rule that gave coal plants a deadline of August 2020 to stop shipping waste to impoundments lacking liners. However, plant operators may be given more time—up to eight years—if regulators can be persuaded that an alternative approach deserves more attention. The new regulation stipulates that coal plants that are retiring by 2028 will not have to dispose of ash into dry landfills or treat wastewater. The EPA estimated that these changes would save the coal industry around $175 million in compliance costs.[27] While the Trump administration contended that it is appropriate to include cost in considering changes to the coal ash rule, others disagreed, suggesting that to do so violates the RCRA's statutory mandate.[28]

State Regulators and CCRs

Historically, coal dust issues have not received the kind of regulatory attention that higher-profile environmental polices like the Clean Air Act or Superfund have received. Coal-producing states often favor high levels of production for both jobs and power generation and are inclined to cede considerable regulatory oversight to utility commissions or energy agencies that are reluctant to challenge companies that violate environmental standards. And at the lower end of the economic scale, states such as Wyoming, Kentucky, and West Virginia are largely dependent on resource extraction since they lack the diversity of employment opportunities found in more affluent producer states such as Colorado, Texas, and Pennsylvania. However, all are potentially subject to change from unforeseen factors like major spills or dam failures, gubernatorial elections, changing attitudes of utility officials, and public concern emerging from new information about nearby water quality risks.

An example drawn from politically conservative Georgia shows how the interplay among these factors can bring about change from the bottom up. In 2016, a rural county newspaper stoked fear among residents when reporters revealed companies were shipping coal ash to the local landfill and groundwater sources had been contaminated. According to Patrik Jonnson, a local groundswell of opposition to these shipments demonstrated a "sense of communal loss due to government malfeasance and disdain for the plight of average Americans."[29] This opposition led to a revocation of the permit for the coal ash dump and to the belief that the community, like others in Georgia, was increasingly sensitive to environmental problems and that greater transparency was needed in communications between regulators and the public.[30]

It is useful to mention the catalytic importance of the EPA's 2015 coal ash rule—notably, the inclusion of requirements for coal plant managers to monitor groundwater quality and to post the information on a public website. The accessibility of this publicly available data made it possible for groups such as the Environmental Integrity Project and the Sierra Club to prepare reports that revealed major groundwater quality problems in coal-producing states. Such information could also become part of a legally justifiable strategy to challenge environmentally questionable policies.[31]

The increased availability of water quality information has led to a willingness among some state lawmakers to become less deferential to the financial and policy concerns of utilities. The political dominance of utilities at the state level historically has been related to dependence on coal as a key fuel for generating electricity and fealty to the importance of a business model that placed greater emphasis on the reliability of energy supplies than on environmental impacts. But this situation has changed. A steep decline in coal production can be attributed to the rising costs of meeting federal environmental quality requirements, while the costs of producing power from cleaner energy sources such as natural gas and renewable energy have gone down. These changes, in turn, have contributed to wider decisional latitude for utility regulators.[32]

While some southern states like Kentucky have largely ignored coal ash problems, others—such as Virginia, South Carolina, and North Carolina—have enacted legislation requiring utilities to close or clean up CCRs.[33] Many rightly view these legislative mandates as a sign of progress in efforts to balance environmental protection with energy production. However, it is important to note that the closure of coal plants does not necessarily lead to a corresponding decline in pollution from coal ash management problems. States often fail to plan for the process of decommissioning a coal plant without considering the costs of closing or relocating coal slurry. According to Daniel Raimi,[34] unexpected costs will ultimately be shouldered by ratepayers, stockholders, or taxpayers.[35]

The Cases: Kentucky and North Carolina

Kentucky

According to the U.S. Energy Information Administration (EIA), Kentucky ranks fifth in the nation's coal production. Not surprisingly, the state remains heavily dependent on coal as a fuel source—75 percent of its electrical generation is derived from coal power plants.[36] But high energy production also results in considerable waste. The industry is responsible for generating nine million tons of slurry and dust (also ranking fifth nationally). Kentucky has nineteen coal plants, forty-three surface impoundments, and twelve landfills that accept waste. Coal ash and slurry waste sites create environmental and safety risks for residents. No other state except Ohio has as many dams classified as "high hazard" (eight), and few of these have been inspected or monitored.[37]

Key regulators in Kentucky include the state's Division of Waste Management within the Kentucky Energy and Environment Cabinet. The waste management division is responsible for the issuance of permits to dispose of coal slurry or dust. Regulated parties—that is, coal plants—are required to keep records of CCRs. While operators are supposed to avoid disposal of by-products near water sources, exceptions may occur if operators can prove that adverse impacts can be avoided.[38] In general, enforcement of environmental requirements receives little emphasis since inspectors are prohibited from adopting stricter procedures than those of the EPA. Efforts are made to recycle substances for beneficial use whenever possible. Otherwise, most disposal of CCRs takes place in landfills.

While the cabinet is responsible for regulating coal policy issues, including the storage or disposal of CCRs, a tendency to hold meetings disproportionately with regulated coal plant operators or their organizational allies, such as the Kentucky Coal Association or Friends of Coal, has generated conflict in recent years. Kentucky Coal Association lobbyists have sided with utility interests, arguing that EPA decisions under the Obama administration were overly aggressive and hurt job creation efforts within the commonwealth.[39] Citizens and environmental groups, such as the Kentucky Resources Council, complained at regulatory hearings about a lack of participatory opportunities.[40]

An important controversy spanning several years reveals procedural issues. The controversy involved a series of backroom meetings between utility officials and former governors Steve Beshear and Matt Bevin. The goal of the meetings was to craft a plan aimed at giving the state's utilities maximum flexibility to self-regulate the storage of CCRs near power plants. These meetings were initially defended by cabinet secretary Charles Snavely as providing ample time for public deliberation, but reporters' uncovered documents suggesting otherwise. The draft regulations prepared in 2015 were altered and weakened significantly before being submitted to the Legislative Research Commission in October 2016 by Governor Bevin.[41]

Suggested changes included doing away with the state's public permitting process for surface impoundments and some landfills based on the argument that they duplicated EPA rules already in place. The final version of the Kentucky Energy and Environment Cabinet's plan mandated that electric utilities provide advance notice to regulators before building new ponds or landfills to store CCRs, but it did not require any sort of preconstruction review process or public input.[42]

This led to a lawsuit by the Kentucky Resources Council on behalf of a landowner whose water well was located near ash ponds. The council challenged the procedural alterations on the basis that participatory opportunities were not provided during the early stages of development planning. The court decision was critical of the cabinet's changes, concluding that they were less likely to protect consumers than to save money for the utilities. The judge also indicated that input into regulatory decisions were largely limited to utility stakeholders rather than a wider range of interests.[43]

This situation suggests that the state's energy and environment cabinet is perhaps hamstrung by a dysfunctional organizational culture (see Chapter 5) that favors decisions maintaining control and secrecy preferred by industry interests above those promoting greater emphasis on environmental health and safety policy goals. An evaluation of state actions has been undertaken by Earthjustice and the Environmental Integrity Project in a report that highlights several deficiencies associated with the state's regulatory policies. Chief among these is the failure to require liners at landfills or surface impoundments to reduce risks to groundwater quality. There is also a lack of state oversight of dams and slurry ponds; few reporting requirements are expected of coal operators, and bond requirements are minimal. Nor is there much in the way of planning for emergency actions for communities that are potentially vulnerable to a high-risk dam breach.[44]

North Carolina

North Carolina is not a major coal-producing state. It relies on nuclear power, and increasingly on natural gas and solar, to generate electricity. The state also relies on a still sizable but declining amount of coal imported from other states—primarily West Virginia, Pennsylvania, and Kentucky—at its fourteen coal-fired power plants.[45] These plants are owned and operated by the state's largest utility, Duke Energy, and are responsible for producing about 5.5 billion tons of CCRs annually.[46] While North Carolina has long been known as a contributor to coal waste production, the state was largely inactive from a policy perspective. Surface impoundments were exempted from regulatory coverage under the Dam Safety Act of 1967. Moreover, these slurry ponds were not required to have protective liners or caps to prevent leaks and did nothing to monitor groundwater quality.

But the situation in North Carolina clearly changed in the aftermath of the Kingston, Tennessee, coal ash spill in 2008, which raised public awareness of the health and environmental risks of waste from coal production. Not only did the EPA take notice by proposing a major new rule dealing with coal ash disposal during Obama's first term, but state lawmakers in North Carolina decided to address the issue. The main catalyst for legislative action was the catastrophic spill that released thirty-nine thousand tons of coal slurry from a Duke Energy coal plant into a seventy-mile stretch of the Dan River in 2014.[47]

Politics and gubernatorial campaigns became intertwined with coal ash regulatory issues shortly thereafter. Pat McCrory, a former employee of Duke Energy and mayor of Charlotte, was elected governor of North Carolina in 2012 and worked with the Republican-controlled legislature to pass the Coal Ash Management Act of 2014. The law called for state regulators to set deadlines for closing coal ash ponds that were leaking or were structurally unstable, especially facilities classified as "high risk" (for loss of life or infrastructure damage) or "intermediate risk" (for economic loss or environmental damage). If implemented expeditiously, the law would have called for the excavation of coal ash, but regulators concluded that the suggested time lines were overly ambitious.[48]

While the Coal Ash Management Act dealt with developing more environmentally responsible cleanup plans generally, another major concern revolved around the question of who pays for expensive ash cleanup and closure actions. Should Duke Energy be allowed to recoup some expenses by raising consumers' energy bills, or should company shareholders be responsible for paying a greater share to "clean the highest risk sites by removing the ash to a lined landfill"?[49] The initial response of North Carolina regulators to the Dan River spill was viewed by some as overly sensitive to utility interests since it included a minimal fine coupled with no requirement that Duke Energy remove the ash from its porous and unlined surface impoundments.[50] In 2016, as the gubernatorial campaign was heating up, the legislature passed a bill to reconstitute a citizens' commission to oversee the Department of Environmental Quality's (DEQ) efforts to regulate coal ash basins around the state and to provide drinking water to residents living in close proximity to Duke's leaking coal plants. This was opposed by Governor McCrory, and the bill was subsequently vetoed as an unwarranted restriction on his authority.[51]

Meanwhile, the Democratic gubernatorial candidate, Roy Cooper, adopted a campaign narrative largely focused on McCrory's overly close ties to Duke Energy, McCrory's former employer. Cooper directed attention to decisions made by a state utilities commission stacked with McCrory appointees to approve Duke's request for rate hikes to pay for ash cleanup costs. Cooper also contended that McCrory politicized science by allowing state toxicologists to mollify consumers about the safety of drinking water despite tests revealing unsafe levels of chromium found in water wells located close to coal ash piles.[52] According to Elizabeth Ouzts, an examination of court depositions revealed the governor called on public health officials to "downplay the risks to water supplies."[53] One consequence of this finding was the subsequent resignation of the state epidemiologist.

Cooper narrowly won the governor's office in 2016 but did not mention coal ash in either his inaugural address or his State of the State address. He paid relatively little lip service to the topic until it became a pollution control issue during another natural catastrophe—Hurricane Florence in 2018. This storm led to considerable flooding in North Carolina, including inland areas. State officials at the DEQ paid close attention to affected facilities, such as a dam breach that resulted in coal ash flowing into the Cape Fear River as well as a power plant near Wilmington, where floodwaters poured though basins containing coal slurry. Complicating matters further, floodwaters exacerbated pollution from not just coal ash but ponds containing hog waste from concentrated animal feeding operations.[54]

In January 2019, public hearings pertaining to the cleanup of coal ash impoundments were held throughout North Carolina. Duke Energy initially argued that a cap-in-place approach would work, but this spurred an overwhelming negative response from newspapers around the state to the company's position. Citizen comments overwhelmingly called for coal ash stored at remaining dam sites to be moved to storage areas with liners. This outcry was followed by an order to DEQ officials from Governor Cooper to review the science relating to a prospective closure date for remaining coal sites. He then promised that the Department of Environmental Quality would follow though to protect affected communities from CCRs. After initially expressing reservations, Duke Energy altered its policy and agreed to complete cleanup and storage for the remaining waste.[55]

Concluding Thoughts

The recognition of coal ash storage and disposal as an important regulatory issue has increased considerably in the early twenty-first century. King coal was viewed as the preeminent fuel source for generating electricity in the United States throughout the twentieth century. However, negative environmental impacts coupled with the rise of renewable energy sources and energy efficiency programs has contributed to a decline in coal use and the phasing out of plants that are obsolete or uneconomical. Specific environmental harms such as those caused by coal ash have become increasingly visible as sources of air and water pollution thanks to events such as the 2008 collapse of a containment pond at the Tennessee Valley Authority's coal power plant in Kingston, Tennessee, as well as the breach of a coal slurry pond at a Duke Energy power station in 2014 that contaminated a major stretch of the Dan River in North Carolina.

These events led to the partial nationalization of coal ash policy, where states had previously controlled this policy space. President Obama's EPA administrator Lisa Jackson proposed a new rule based on the solid waste disposal clause imbedded in the Resource Conservation and Recovery Act, and the new rule was eventually adopted in 2015. While environmentalists were disappointed that a tougher hazardous waste classification was not chosen, the rule did establish safety standards to aid states in determining appropriate levels of structural integrity for dams and surface impoundments. It also mandated that utilities and dam operators require groundwater testing for ash ponds as well as liners to prevent leaks. Testing results were expected to be published at publicly accessible websites.

The latter change is potentially quite important in terms of information disclosure. Publicly available information about state-level regulatory risks has been used by groups such as Earthjustice and the Environmental Integrity Project to write a series of reports that provide a scientific foundation for litigation against violators of coal ash policies. In addition, the disclosure of environmental harms linked to the failure of coal ash policies can facilitate local or state enforcement efforts as well as efforts to mobilize public concerns about water quality.

Because the EPA was not given an explicit enforcement role, the agency was assigned the responsibility of working with state officials to design a permitting process. Time lines were also established to determine when coal ash waste should be transported from higher-risk dams or ponds to lined landfills or storage areas. When the Trump administration took over, regulatory expectations were relaxed, often giving state regulators more time to meet deadlines. The role of the courts as a policy-making venue has provided environmental interests with one pathway to resist unwanted changes. Initiating lawsuits based on RCRA statutory requirements or the disclosure of environmentally worrisome information has made it easier to challenge attempts to roll back or weaken coal ash management policies.

While a regulatory policy goal of the EPA's new responsibilities was to better harmonize a disparate set of state policies, considerable variation remains in play. The Kentucky and North Carolina cases suggest that contextual factors, such as each state's relative dependence on coal and the diversity of energy options, tend to make a difference. Kentucky is a poor state that remains dependent on coal as a source of fuel and jobs, while North Carolina has a more diversified economy and other power sources besides coal—notably, natural gas and nuclear power.

Both cases also reveal a key role for governors to straddle the line between energy production and environmental protection. This chapter suggests that a change in chief executives can result in different outcomes at both the federal level (EPA) and within states (North Carolina). The Tar Heel State certainly reveals differences between Republican governor McCrory and Democratic governor Cooper in holding utility officials accountable for mishandling coal ash closure and containment policies. As noted in Chapter 6, gubernatorial succession policies can also influence the enforcement of oil and gas regulations, even when both governors represent the same political party. We now turn to our final chapter to explore options for future environmental policy.

Notes

1. U.S. Energy Information Administration, "Coal Explained: Coal and the Environment," updated December 1, 2020, https://www.eia.gov/energyexplained /coal/coal-and-the-environment.php.

2. Climate Nexus, "What's Driving the Decline of Coal in the United States?" March 21, 2019, https://climatenexus.org/climate-issues/energy/whats-driving -the-decline-of-coal-in-the-united-states/.

3. Shalina Chatlani, "Two Years after EPA's Coal Ash Rule, Progress Depends on States," *Utility Dive*, May 24, 2016.

4. Elmer Gilmer, "Coal Ash: Disposal Sites Aren't Covered by Clean Water Act—Court," *Greenwire*, September 24, 2018.

5. Emmarie Huetteman, "Coal Ash to Protect Water Supply," *New York Times*, December 19, 2014.

6. Duke Energy, "Coal Ash Recycling," 2018, http://duke-energy.com/ash -management.

7. Elizabeth Connors, "Coal-Ash Management by U.S. Electric Utilities: Overview and Recent Developments," *Utilities Policy* 34 (2015): 30–33.

8. A. Dennis Lemly and Joseph P. Skorupa, "Wildlife and the Coal Waste Policy Debate: Proposed Rules for Coal Waste Disposal Ignore Lessons from 45 Years of Wildlife Poisoning," *Environmental Science and Technology* 46, no. 16 (2012), 8595–8600.

9. Abel Russ, Courtney Bernhardt, and Lisa Evans, "Coal's Poisonous Legacy: Groundwater Poisoned by Coal Ash across the U.S.," Environmental Integrity Project, Report, March 4, 2019; revised July 11, 2019, 13.

10. James Bruggers, "Coal Ash Contaminates Groundwater at 91% of U.S. Coal Plants, Tests Show," *Inside Climate News*, March 4, 2019.

11. Ibid., 4.

12. Connors, "Coal Ash Management," 32.

13. Caitlyn McCoy and Laura Bloomer, "Coal Ash Rule," Harvard Environmental and Energy Law Program, November 5, 2019, https://eelp.law. harvard .edu/2017/12/coal-ash-rule/.

14. Jonathan M. Fisk, A. J. Good, and Steven Nelson, "Collaboration after Disaster: Explaining Intergovernmental Collaboration during the EPA Gold King Mine and TVA Coal Ash Recoveries," *Risk, Hazards and Crisis in Public Policy* 10, no. 1 (2019): 52–73.

15. U.S. Environmental Protection Agency, "Fact Sheet: Coal Combustion Residuals Surface Impoundments with High Hazard Potential Ratings," EPA530 -F-09-006, June 2009, https://nepis.epa.gov/Exe/ZyPDF.cgi/ P10048EX.PDF ?Dockey=P10048EX.PDF.

16. Huetteman, "Coal Ash to Protect Water Supply."

17. Katherine Clements, "The Coal Ash Rule Trilogy Spanning Obama, Trump, and the D.C. Circuit," Harvard Environmental and Energy Law Program, January 28, 2020, 2.

18. Russ, Bernhardt, and Evans, "Coal's Poisonous Legacy," 38.

19. Joby Warrick, "Coal-Ash Dumps to Face Restrictions Six Years after Billion-Gallon Spill in Tennessee," *Washington Post*, December 19, 2014.

20. Congressional Research Service, "State Programs for 'Coal Ash' Disposal in the WIIN Act," Report no. IN10585, December 8, 2016, https://www .everycrsreport.com/ reports/IN10585.html.

21. Edward Weber, David Bernell, Hilary Boudet, and Patricia Fernandez-Guajardo, "Energy Policy: Fracking, Coal, and the Water-Energy Nexus," in *Environmental Policy: New Directions for the Twenty-First Century*, 10th ed., ed. Norman Vig and Michael Kraft (Washington, DC: CQ Press, 2019), 206.

22. Nadia Popovich, Livia Albeck-Ripka, and Kendra Pierre-Louis, "95 Environmental Rules Being Rolled Back under Trump," *New York Times*, December 21, 2019.

23. Cited in Sean Reilly, "Coal Ash: EPA Offers Guidance to States That Want to Manage Permitting," *E&E News PM*, August 10, 2017.

24. Catherine Morehouse, "EPA Makes Georgia 2nd State to Operate Coal Ash Program, Proposes 'Efficient' Rule for Other 48," *Utility Dive*, December 23, 2019.

25. McCoy and Bloomer, *Coal Ash Rule*, 5.

26. Ibid., 6.

27. Ariel Wittenberg and Sean Reilly, "EPA Punches Loopholes in Coal Ash Regs," *Greenwire*, November 4, 2019.

28. Clements, "The Coal Ash Rule Trilogy," 4.

29. Patrik Jonnson, "Why Business-Friendly Georgia Got Tough on Environmental Regulation," *Christian Science Monitor*, November 21, 2019.

30. Ibid., 4.

31. Steven Mufson and Brady Dennis, "Report Finds Widespread Contamination at Nation's Coal Ash Sites," *Washington Post*, March 4, 2019.

32. Climate Nexus, "What's Driving the Decline of Coal," 1.

33. Max Blau, "Kentucky among Southern States Split on Dealing with EPA's Rollback of Coal Ash Disposal Regulations," *Stateline*, December 12, 2019. See also Frank Holleman, "Cap Coal Ash in Place? Duke and Others Have Learned Better," *Utility Dive*, February 24, 2020.

34. Daniel Raimi, "Decommissioning US Power Plants: Decisions, Costs, and Key Issues," *Resources for the Future*, October, 2017, 45.

35. Ibid.

36. U.S. Energy Information Administration, "Kentucky: State Profile and Energy Estimates," updated June 18, 2020, https://www.eia.gov/state/analysis .php?sid=KY.

37. Kate Carpenter, "States in Need of Repair: The Interaction of Coal Ash with the Human Environment and the Effectiveness of Statutes in Kentucky and Ohio," *Journal of Animal and Environmental Law* 6 (2014): 123.

38. Ibid., 36.

39. Blau, "Kentucky among Southern States," 2.

40. Erica Peterson, "Kentucky Regulators, Industry Reps Privately Rewrote Coal Ash Rules," *Environmental Investigations*, January 17, 2017, 2.

41. Morgan Watkins, "Vote Delayed on Bevin Plan for Coal Ash Dumps," *Louisville Courier-Journal*, February 10, 2017.

42. James Bruggers, "Kentucky Seeks to Slash Coal Ash Oversight," *Louisville Courier-Journal*, December 8, 2016.

43. Robert Walton, "Kentucky Circuit Court Overturns Coal Ash Rules," *Utility Dive*, February 1, 2018.

44. Lisa Evans, Michael Becher, and Bridget Lee, "State of Failure: How States Fail to Protect Our Health and Drinking Water from Toxic Coal Ash," Environmental Integrity Project and the Appalachian Mountain Advocates, 2013, 15.

45. U.S. Energy Information Administration, "North Carolina: State Profile and Energy Estimates," updated November 19, 2020.

46. Evans, Becher, and Lee, "State of Failure," 16–17.

47. Andrew Kenney, "NC Lawmakers Pass Coal Ash Legislation," *Raleigh News and Observer*, August 20, 2014.

48. Chatlani, "Two Years," 3–5.

49. Kenney, "NC Lawmakers," 1.

50. Trip Gabriel, "Ash Spill Shows How Watchdog Was Defanged," *New York Times*, February 28, 2014.

51. Craig Jarvis, "Deconstructing Latest NC Coal Ash Controversy," *Raleigh News and Observer*, May 28, 2016.

52. Bruce Henderson, "McCrory's Duke Energy Ties, and Coal Ash Response, Become a Campaign Issue," *Charlotte Observer*, October 21, 2016.

53. Elizabeth Ouzts, "After Campaign Pledge, N.C. Gov. Roy Cooper Faces Test on Coal Ash," *Energy News Network*, January 24, 2019.

54. Glenn Thrush and Kendra Pierre-Louis, "Florence's Floodwaters Breach Defenses at Duke Energy Plant, Sending Toxic Coal Ash into River," *New York Times*, September 21, 2018.

55. Frank Holleman, "Cap Coal Ash in Place? Duke and Others Have Learned Better," *Utility Dive*, February 24, 2020.

Suggested Readings

Clements, Katherine. The Coal Ash Rule Trilogy Spanning Obama, Trump, and the D.C. Circuit. Harvard Environmental and Energy Law Program. January 28, 2020. Available at https://eelp.law.harvard.edu/2020/01/the-coal-ash-rule-trilogy-spanning-obama-trump-and-the-d-c-circuit/.

Connors, Elizabeth. "Coal-Ash Management by U.S. Electric Utilities: Overview and Recent Developments." *Utilities Policy* 34 (2015): 30–33.

Demmerle, Amanda P. "Pain in the Ash: How Coal-Fired Power Plants Are Polluting Our Nation's Waters Without Consequences." *West Virginia Law Review* 122 (2019): 289.

Fisk, Jonathan M., A. J. Good, and Steven Nelson. "Collaboration after Disaster: Explaining Intergovernmental Collaboration during the EPA Gold King Mine and TVA Coal Ash Recoveries." *Risk, Hazards and Crisis in Public Policy* 10, no. 1 (2019): 52–73.

McGarity, Thomas O. *Pollution, Politics, and Power: The Struggle for Sustainable Electricity*. Cambridge, MA: Harvard University Press, 2019.

Bibliography

Blau, Max. "Kentucky among Southern States Split on Dealing with EPA's Rollback of Coal Ash Disposal Regulations." *Stateline*, December 12, 2019.

Bruggers, James. "Coal Ash Contaminates Groundwater at 91% of U.S. Coal Plants, Tests Show." *Inside Climate News*, March 4, 2019.

———. "Coal Ash Contaminating Groundwater in at Least 22 States, Utility Reports Show." *Inside Climate News*, January 18, 2019.

———. "Kentucky Seeks to Slash Coal Ash Oversight." *Louisville Courier-Journal*, December 8, 2016.

Carpenter, Kate. "States in Need of Repair: The Interaction of Coal Ash with the Human Environment and the Effectiveness of Statutes in Kentucky and Ohio." *Journal of Animal and Environmental Law* 6 (2014): 123.

Chatlani, Shalina. "Two Years after EPA's Coal Ash Rule, Progress Depends on States." *Utility Dive*, May 24, 2016.

Clements, Katherine. "The Coal Ash Rule Trilogy Spanning Obama, Trump, and the D.C. Circuit." Harvard Environmental and Energy Law Program.

January 28, 2020. Available at https://eelp.law.harvard.edu/2020/01/the-coal
-ash-rule-trilogy-spanning-obama-trump-and-the-d-c-circuit/.

Climate Nexus. "What's Driving the Decline of Coal in the United States?"
March 21, 2019. Available at https://climatenexus.org/climate-issues/energy
/whats-driving-the-decline-of-coal-in-the-united-states/.

Congressional Research Service. *State Programs for "Coal Ash" Disposal in the
WIIN Act*. Report no. IN10585. December 8, 2016. Available at https://www
.everycrsreport.com/reports/IN10585.html.

Connors, Elizabeth. "Coal-Ash Management by U.S. Electric Utilities: Overview
and Recent Developments." *Utilities Policy* 34 (2015): 30–33.

Daniels, John. "Coal Ash and Groundwater: Past, Present and Future Implications
of Regulation." *William & Mary Environmental Law and Policy Review* 40,
no. 6 (2016). Available at https://scholarship.law.wm.edu/wmelpr/vol40/
iss2/6.

Duke Energy. "Coal Ash Recycling: Improving Our Communities, the Environment
and the Economy by Putting Ash to Use." 2018. Available at www.duke-energy
.com/ash-management.

Evans, Lisa, Michael Becher, and Bridget Lee. "State of Failure: How States Fail to
Protect Our Health and Drinking Water from Toxic Coal Ash." Environmental
Integrity Project and the Appalachian Mountain Advocates. 2013. Available
at https://earthjustice.org/sites/ default/files/StateofFailure_2013-04-05.pdf.

Fisk, Jonathan M., A. J. Good, and Steven Nelson. "Collaboration after Disaster:
Explaining Intergovernmental Collaboration during the EPA Gold King
Mine and TVA Coal Ash Recoveries." *Risk, Hazards and Crisis in Public
Policy* 10, no. 1 (2019): 52–73.

Gabriel, Trip. "Ash Spill Shows How Watchdog Was Defanged." *New York Times*,
February 28, 2014.

Gilmer, Elmer. "Coal Ash: Disposal Sites Aren't Covered by Clean Water Act—
Court." *Greenwire*, September 24, 2018.

Henderson, Bruce. "McCrory's Duke Energy Ties, and Coal Ash Response,
Become a Campaign Issue." *Charlotte Observer*, October 21, 2016.

Holleman, Frank. "Cap Coal Ash in Place? Duke and Others Have Learned
Better." *Utility Dive*, February 24, 2020.

———. "Pressure from Citizens in Coal Ash Communities Won Cleanup
Settlement." *WRAL Evening News*, January 12, 2020. Available at https://www
.wral.com/frank-holleman-pressure-from-citizens-in-coal-ash-communities
-won-cleanup-settlement/18879317/.

Huetteman, Emmarie. "Coal Ash to Protect Water Supply." *New York Times*,
December 19, 2014.

Jarvis, Craig. "Deconstructing Latest NC Coal Ash Controversy." *Raleigh News
and Observer*, May 28, 2016.

Jonnson, Patril. "Why Business-Friendly Georgia Got Tough on Environmental
Regulation." *Christian Science Monitor*, November 21, 2019.

Kenney, Andrew. "NC Lawmakers Pass Coal Ash Legislation." *Raleigh News and Observer*, August 20, 2014.

Lemly, A. Dennis, and Joseph P. Skorupa. "Wildlife and the Coal Waste Policy Debate: Proposed Rules for Coal Waste Disposal Ignore Lessons from 45 Years of Wildlife Poisoning." *Environmental Science and Technology* 46, no. 16 (2012): 8595–8600.

McCoy, Caitlyn, and Laura Bloomer. "Coal Ash Rule." Harvard Environmental and Energy Law Program. November 5, 2019. Available at https://eelp.law .harvard.edu/2017/12/coal-ash-rule/.

Morehouse, Catherine. "EPA Makes Georgia 2nd State to Operate Coal Ash Program, Proposes 'Efficient' Rule for Other 48." *Utility Dive*, December 23, 2019.

———. "Georgia Approves Almost $1.8 Billion in Rate Hikes to Cover Coal Ash Cleanup." *Utility Dive*, December 18, 2019.

Mufson, Steven, and Brady Dennis. "Report Finds Widespread Contamination at Nation's Coal Ash Sites." *Washington Post*, March 4, 2019.

Peterson, Erica. "Kentucky Regulators, Industry Reps Privately Rewrote Coal Ash Rules." *Environmental Investigations*, January 17, 2017.

Popovich, Nadia, Livia Albeck-Ripka, and Kendra Pierre-Louis. "95 Environmental Rules Being Rolled Back under Trump." *New York Times*, December 21, 2019.

Raimi, Daniel. "Decommissioning US Power Plants: Decisions, Costs, and Key Issues." *Resources for the Future*, October 2017.

Reilly, Sean. "Coal Ash: EPA Offers Guidance to States That Want to Manage Permitting." *E&E News PM*, August 10, 2017.

Russ, Abel, Courtney Bernhardt, and Lisa Evans. "Coal's Poisonous Legacy: Groundwater Poisoned by Coal Ash across the U.S." Report. Environmental Integrity Project. March 4, 2019; revised July 11, 2019.

Thrush, Glenn, and Kendra Pierre-Louis. "Florence's Floodwaters Breach Defenses at Duke Energy Plant, Sending Toxic Coal Ash into River." *New York Times*, September 21, 2018.

U.S. Energy Information Administration. "Coal Explained: Coal and the Environment." Updated December 1, 2020. Available at https://www.eia.gov/energy explained/coal/coal-and-the-environment.php.

———. "Kentucky: State Profile and Energy Estimates." Updated June 18, 2020. Available at https://www.eia.gov/state/analysis.php?sid=KY.

———. "North Carolina: State Profile and Energy Estimates." Updated November 19, 2020. Available at https://www.eia.gov/state/analysis.php?sid=NC.

U.S. Environmental Protection Agency. "Coal Ash Basics." Available at https:// www.epa.gov/coalash/coal-ash-basics.

———. "Fact Sheet: Coal Combustion Residuals (CCR) Surface Impoundments with High Hazard Potential Rating." EPA530-F-09-006. June 2009. Available at https://nepis.epa.gov/Exe/ZyPDF.cgi/P10048EX.PDF?Dockey=P10048EX .PDF.

———. "Permit Programs for Coal Combustion Residual Disposal Units." Available at https://www.epa.gov/coalash/permit-programs-coal-combustion-residual-disposal-units.

Walton, Robert. "Kentucky Circuit Court Overturns Coal Ash Rules." *Utility Dive*, February 1, 2018.

Warrick, Joby. "Coal-Ash Dumps to Face Restrictions Six Years after Billion-Gallon Spill in Tennessee." *Washington Post*, December 19, 2014.

Watkins, Morgan. "Vote Delayed on Bevin Plan for Coal Ash Dumps." *Louisville Courier-Journal*, February 10, 2017.

Weber, Edward, David Bernell, Hilary Boudet, and Patricia Fernandez-Guajardo. "Energy Policy: Fracking, Coal, and the Water-Energy Nexus." In *Environmental Policy: New Directions for the Twenty-First Century*, 10th ed., edited by Norman Vig and Michael Kraft, 194–218. Washington, DC: CQ Press, 2019.

Wittenberg, Ariel. "Republicans, States Alarmed about EPA Action on Groundwater." *E&E News Daily*, April 19, 2018.

Wittenberg, Ariel, and Sean Reilly. "EPA Punches Loopholes in Coal Ash Regs." *Greenwire*, November 4, 2019.

8

Our Environmental Policy Tomorrow

SARA R. RINFRET

We have now come full circle in our regulatory journey, exploring who makes U.S. environmental policy through the lens of regulatory processes. However, as we conclude *Who Really Makes Environmental Policy?* we note that 2020–2021 was a year for the history books. The COVID-19 global pandemic forced us to shelter in place, not leaving our homes for months. As a result, people did not drive their cars, oil prices dropped, and the U.S. Environmental Protection Agency (EPA) offered enforcement discretion.

The uncertainty of a global pandemic, I argue, presents an opportunity for environmental policy. It encourages us to pause, reflect, and reimagine how we live and work. For example, Tamarra Thiesen of *Forbes* magazine argues that COVID-19 brings "the planet a much needed breath of fresh air."[1] And experts in their respective fields are in the limelight as politicians seek solutions.[2] COVID-19 serves as an important reminder about the theme of this text—illustrating how individuals outside the halls of Congress develop solutions for public sector problems.

This final chapter brings together the regulatory expertise documented throughout this book to offer a new direction for U.S. environmental policy. Now more than ever, the expertise of the front-line regulatory actors of environmental policy—the rule writers and inspectors—will shepherd the United States into our environmental tomorrow. As Sara Rinfret, Denise Scheberle, and Michelle Pautz note, "These individuals can serve as consensus makers in an era of policy divisiveness."[3]

This final chapter begins with a brief recap of the connections across chapters to illustrate how information sharing, technical expertise, and reconciling differences collectively capture the perspectives across the book's case studies. This summary serves as the foundation for a descriptive regulatory framework—environmental policy tomorrow.

Why Environmental Regulations Matter

Throughout this edited volume, we illustrate where environmental policy is made—within federal, state, and local environmental agencies. We have seen how rule writers and inspectors work to implement vague congressional statutes to ensure that bike riding in a national park does not impact sensitive plant life or species, and how best to deal with oil and gas quagmires in Colorado. A holistic review illustrates broader themes across the chapters, which serve as a foundation for the nation's environmental tomorrow.

Chapter 1 serves as a guidepost defining the intricacies of U.S. environmental regulations, from congressional delegation of authority to defining the stages of the rulemaking process. Some of the rationales are explained for why Congress delegates its authority to administrative agencies to make policy (e.g., lack of expertise or time). This delegation of authority is monitored by Congress through the Administrative Procedure Act (APA) of 1946. Nonetheless, agencies can and do make policy through administrative rulemaking. This process occurs over several stages, ranging from rule development, notice and comment, and rule finalization to environmental inspections by states. The stages of rulemaking, unfamiliar to many, form a participatory process for federal and state agencies to engage the public in information sharing to assist in improving the quality of environmental policy more broadly.

At the heart of rulemaking is participation. Chapter 2 unveils how agencies begin the information sharing process by meeting with potentially affected stakeholders. The exchange of ideas among industry, academics, area experts, and rule writers is instrumental in the creation of an agency rule. Chapter 2's exploration of the dog management rule in Golden Gate National Recreation Area makes clear that without expertise from a myriad of actors, the rule would never have been proposed or finalized. It is also important to remember that the pre-rule stage of rulemaking is not required by the APA to be part of the record. Instead, it allows for an agency to work with stakeholder groups to develop the language of the rule. The pre-rule stage can take years or even decades, as did the long-standing issues of dog walking in Golden Gate National Recreation Area.

Chapter 2's details of rule development transition to Chapter 3's explanation of how and why individuals submit a public comment. For some individuals or groups, submitting a public comment allows them to formally indicate their preference for a given rule. As indicated in Chapter 3, the individuals most impacted by the rule (e.g., business interests) more actively participate in public commenting. Yet the evolution of the commenting process has been transformed by e-rulemaking, offering positives and pitfalls, from increased access to bots and fake comments.

As Chapter 4 suggests, just because a rule is final does not mean the process is over. The president and Congress can still make an impact through executive orders and political appointments of federal agency administrators. Chapter 4 additionally serves as a reminder about the role of state and local actors, setting the tone for Chapters 5, 6, and 7.

Chapters 5, 6, and 7 offer a glimpse into where the rubber meets the road and how environmental rules are implemented at the state and local level. Chapter 5, for example, documents how state actors such as environmental inspectors should be classified not as cops but as heroes who ensure compliance with environmental law. Chapter 6 explores the interconnectedness between state inspectors and rulemaking processes, using oil and gas regulation in Colorado as an example. Chapter 7 illustrates how state and federal actors work together (or fail to do so), revealing the complexities of regulating energy production and environmental protection.

Each chapter highlights where environmental policy is made in the United States—in federal and state environmental agencies. These explanations are made possible through rich chapter case studies. Collectively, the case studies offer three important themes: information sharing, technical substance, and reconciliation of differences.

Information sharing. Across the chapter case studies, the sharing of information is a consistent theme about how to create a rule or ensure compliance with an environmental law. For example, rule writers consistently engage with various groups to develop a rule (e.g., dog management, mountain biking, the Tailoring Rule). Post-rulemaking, inspectors actively engage with regulated facilities during the inspection process. Recall the inspector's monitoring of air pollution by a fiberglass manufacturing facility.

As the case studies consistently suggest, environmental regulators, in collaboration with regulated communities, work to ensure standards are met. Sharing information with greater involvement of affected entities allows regulators to understand how goals and objectives can be achieved. As Michelle Pautz and Sara Rinfret suggest, enormous potential exists for knowledge development and information sharing when there is more collaboration throughout the rulemaking process. As a result, what regulators learn with one facility may be easily shared with others, with the ultimate result of better environmental outcomes all around.[4]

Technical expertise. The technical nature of environmental regulations has not gone unnoticed in the chapter case examples. Consider the implementation of the Endangered Species Act (Chapter 4) to oil and gas regulations in Colorado (Chapter 6). It is not members of Congress who determine the amount of acreage needed for the rebound of the northern Rocky Mountain wolf. Rather, it is regulators' extensive interdisciplinary backgrounds (e.g., in science, policy, and law) that provide agencies with the necessary expertise to translate vague legislation into environmental policy. As Sara Rinfret and Michelle Pautz reiterate, "Many of our environmental professionals are trained in specific areas of environmental concerns. For example, environmental inspectors, often, have backgrounds in particular areas of the natural sciences or engineering."[5] Denise Scheberle even goes further to suggest that regulatory professionals, through their technical expertise, "bring wisdom to their actions and employ facts and the best available science to present their case."[6] The broad and diverse technical backgrounds of rule writers and inspectors on the front lines of environmental policy are essential for effectively understanding and crafting policy solutions for environmental policy.

Reconciling differences. Influenced by Harold Lasswell, political scientists often ask "who gets what, when, and how."[7] As demonstrated by the case studies in this volume, stakeholders attempt to exert their influence across rulemaking processes or during regulatory interactions with inspectors. However, influence in environmental regulation is not like writing a campaign check or lobbying a member of Congress.

Although attempts to influence environmental regulatory policy outcomes abound, many actors try to influence rulemaking at different stages of the process. At the center of assessing regulatory influence is the recognition of agencies attempting to reconcile differences, acting as arbiters of influence. Put simply, environmental regulators are experts within a climate of competing interests (environmentalists, industry), each pursuing its own objectives.[8] As a result, environmental regulators merge ideas and reconcile differences to implement environmental policy. Resolving differences is not an easy endeavor, as the case studies suggest, but regulatory experts serve as brokers to find satisfactory resolutions.

Although this recap of the examples in this edited volume is not exhaustive, the purpose here is simple—to reflect and remember that environmental policy would not be possible without implementation by experts. These themes, brought to light in this book's contributions, provide lessons and direction for the volume's descriptive framework—environmental policy tomorrow. As suggested at the beginning of this chapter, COVID-19 allows us to pause and reimagine how we live and work in our environmental policy tomorrow.

Environmental Policy Tomorrow

The origins of environmental policy tomorrow come from the key actors who define the U.S. regulatory apparatus—the rule writers and inspectors on the front lines of environmental policy. For the nation to continue to advance the important role of environmental protection, I suggest that the environmental policy tomorrow framework leverage three Ls: (1) lessons from the past, (2) listening as an engagement tool, and (3) leading with regulatory innovation. On the surface, the criteria for an environmental policy tomorrow may appear commonsensical, but I suggest otherwise. This descriptive framework captures how environmental policy continues to be made through the bureaucracy. I explain each of the three characteristics in turn.

Lessons from the Past

Learning is an iterative process, guided by previous experiences, and it is at the crux of U.S. environmental regulations. The 1970s saw a "remarkable burst" of federal environmental legislation adopted by bipartisan congressional support and Republican president Richard Nixon.[9] This legislation serves as the nation's regulatory infrastructure—a command and control approach. This regulatory apparatus was designed to reduce pollution, save U.S. waterways, and address hazardous waste, but it was not easy. As stated in Chapter 4, environmental regulators were seen as police, pitting regulatory experts against those they regulated.[10]

Over the course of several decades, personnel within the EPA, as just one example, have implemented environmental legislation that protects the nation's air and water. However, this legislation does not come easy, because building collaborative relationships between environmental groups and industry is a difficult endeavor. Environmental inspectors have adapted their regulatory style over time, learning from their experiences in working with regulated facilities (see Chapter 5). Historically, scholars suggested that regulatory enforcement styles occur in a binary fashion—strict or relaxed. However, as inspectors have learned over time, both approaches may be adopted during a site inspection, and different circumstances may warrant different behavior. Thus, regulators who embrace a mixed enforcement style combine various tactics and tools as they seek compliance from the regulated community.[11]

Lessons also exist from shared experiences within federal, state, and local environmental agencies. In 2020, the vast majority of environmental agencies are multigenerational. Drew Calvert suggests, "Today's workforce is more generationally diverse than ever before. With more professionals delaying retirement—either by choice or necessity—organizations now have employees from as many as four different generations working side by side. Soon, for the first time in history, a fifth generation will join the mix."[12] This organizational makeup creates an environment of shared learning. For example, the EPA pairs seasoned rule writers with recent hires so that the latter can better understand the processes involved in engaging with the public, examining public comments, and writing rules.[13] As a result, newly hired staff assist in the advancement of new ideas and knowledge for an agency to consider for adoption. The National Park Service, by contrast, offers a fundamentals training program for new and midlevel employees and senior professionals to help all understand the agency's core values.[14]

Nevertheless, as is seen consistently in all the chapters in this volume, environmental regulators are a community of learners, adapting their approaches to forge a next generation of environmental regulation that embraces external and internal collaboration. Rule writers and inspectors use their expertise to adapt to the ever-changing dynamics of internal and external environments. Fundamental to environmental regulators' adaptability over time is their ability to use listening as a tool for engagement.

Listening as an Engagement Tool

Listening is defined by what environmental regulators most often do—capture public sentiment through a myriad of mechanisms for public participation. Chapter 1 reminds us that listening is central to environmental regulations. When Congress delegates its authority to administrative agencies, the APA requires agencies to provide notice and a space for the public to comment on a notice of proposed rulemaking.

Inevitably, the rules of engagement change over time, adapting to societal needs. Traditionally, federal agencies invited participation to provide feedback for a rule by posting a notice in the local newspaper. In turn, individuals could send comments through the mail (see Chapter 3). Agencies have found ways to actively engage the public through public hearings, electronic submission of public comments, and regulatory negotiations. Depending on the policy area, the listening can occur for years, much like the case of dog management in Golden Gate National Recreation Area.

Agencies rely on listening to fact-based evidence from a wide variety of individuals to better inform the language of a rule. For example, Chapter 3's discussion of the EPA's Tailoring Rule explains how the best available technology was coupled with scientific data to provide a new rule. Although opponents to new regulations always exist, agencies rely on information others might offer because they may be missing to strengthen a given rule. This approach actively encourages participation across all stages of the rulemaking process.

Counter to substantive, fact-driven input, some approach participation in rulemaking through the massive submission of duplicative comments (e.g., an interest group asks its members to submit the same letter). This approach can be employed to lobby a member of Congress, but it is ineffective. Although agencies take record of mass comments submitted, most look instead for substantive information to better inform the rule. The question becomes who and which groups an agency listens to in determining a final rule. According to Marissa Golden, businesses often participate in notice and comment submission and therefore tend to have greater impact on the language of the rule.[15] The participation of businesses, however, does not preclude the public from participating in the process.

In short, public confidence is necessary for environmental regulations to succeed. Environmental regulatory agencies serve as active listeners to engage the public throughout all stages of the rulemaking process. Rule writers serve as consensus builders and brokers among stakeholder groups because they create an open dialogue between agency staff and affected groups of a particular rule. As a result, sharing information can build trust between the agency and stakeholder groups.[16]

Listening is a powerful tool that enables regulatory policy makers to leverage what they have heard into action. Collectively, lessons from the regulatory past and listening approaches assist in broader efforts to lead in regulatory innovations for environmental policy tomorrow.

Leading with Regulatory Innovation

What has been left unsaid in this book is that behind the scenes, career civil servants in environmental agencies are quietly advancing what they do on a day-to-day basis as leaders in environmental regulation. Leading, by definition, engenders action. Arguably, regulatory experts are leaders in finding creative ways to implement environmental regulations or even prevent cheating. We explore a few examples here.

Satellites, remote sensing, drones, and real-time monitoring and reporting are now commonplace approaches for environmental inspectors.[17] Such approaches enable real-time air pollution tracking or water monitoring with the ability to share data.[18] For example, satellites provide some of the most accurate tracking of nitrogen dioxide and particulate matter by observing atmospheric changes from space. Additionally, drones are used to fly above sites (e.g., a water facility) to monitor and obtain data. The data are easily downloadable and can be shared across states for comparison to track monitoring over time.

Smartphone applications have improved the efficiency of conducting site inspections. For years, the Pennsylvania Department of Environmental Quality conducted oil and gas inspections with a paper form. With the creation of a new app, preloaded site visit materials were loaded onto a tablet. Inspectors who use the app on site visits have increased their productivity by almost 29 percent.[19] Many inspectors reported that the devices modernized the process by allowing them to conduct more site visits and ensure environmental protection.

Regulatory technology also serves an important role in meeting compliance goals and even preventing cheating. EPA and California regulatory monitoring exposed Volkswagen's diesel testing scandal. In 2013, regulators were concerned that Volkswagen was exploiting a testing loophole—discrepancies in reporting from test results and real-world performance. In regulatory testing, the Volkswagen Jetta was found to run more cleanly when it was cold instead of warm. With consistent evidence, Volkswagen eventually admitted that it had installed software to detect when a car was being tested and lower its emission levels, thus cheating the system.[20]

A final example is the revitalization of Regulations.gov, the main portal for electronically submitted public comments for federal rules. The website is being modernized to become a more responsive design for tablet, phone, or computer. More importantly, electronic feedback makes it easier for individuals to participate in the process. The goal is to provide a user-friendly site to encourage participation among the general public.[21]

Although these regulatory innovations serve an essential purpose, COVID-19 encourages additional considerations from health care professionals. For areas inaccessible for remote sensing or drone, another option is the use of telepresence robots. Commonly used in medical communities to conduct site visits in rural communities, telepresence robots meet with patients. For example, if a doctor in a rural community cannot meet with a patient, a telepresence robot can be used to discuss issues. Additionally, telepresence robots are utilized for in-home care so a nurse can check in with patients. The same technology could be applied to an inspector and facility, much the same way it helps doctors meet the needs of rural communities.

In summary, learning, listening, and leading—elements of environmental policy tomorrow—remind us to revisit the past and realize the tremendous amount of work environmental regulators on the front lines do to continue to shape environmental policy on a daily basis. Without them, we would not have an environmental policy tomorrow. It is because of their expertise that we continue to learn from the past, engage participants in the process, and find new, innovative processes to continue well into tomorrow.

Concluding Thoughts

It is easy for us to say environmental policy is dead. Why? We read in the news about former president Trump's regulatory rollbacks or how the EPA issued self-monitoring for compliance due to COVID-19, relinquishing its duty to carry out its mission to protect public health.[22] These rollbacks can be viewed with National Geographic's rollback tracker (available at https://www.nationalgeographic.com/news/2017/03/how-trump-is-changing-science-environment/). Despite these significant impacts, the Biden administration, within his first 100 days signals efforts to place environmental policy at the forefront of his agenda.

As suggested at the beginning of this chapter, and throughout this book, environmental regulations invite opportunity. It is important to remember that the origins of U.S. regulatory processes were brought forth by the Great Depression, the New Deal, and World War II. Specifically, the Administrative Procedure Act of 1946 structured rulemaking as a mechanism to ensure the government continued to be adaptable, flexible, durable, and responsive. Environmental regulators will continue to weather storms like COVID-19, clearing pathways for environmental policy tomorrow.

Teen Vogue's Liat Olenick and Alessandro Dal Bon suggest that now, more than ever, is the time to act. For Olenick and Dal Bon, "The world will not, and should not, return to normal. By diving into the uncomfortable questions and opening new doors for unlikely collaborations, above all we hope to open up a space where we can come together to imagine and fight for a better future." Further, "With the world economy in recession, our elected leaders have a chance to restart our economy with a green stimulus. Instead of bailing out the fossil fuel industry, we must bail out workers, invest in renewable energy, infrastructure, and housing, and lay the groundwork for a transformational Green New Deal that will center the experiences of communities on the front-lines of COVID-19 and the climate crisis."[23] These statements, although directed at elected lawmakers, present a window of opportunity for elected (Congress) and nonelected (bureaucracy) institutions to act swiftly together.

Amid negative perceptions that they are lazy bureaucrats, the experts on the front lines of environmental policy emerge as the heroes for environmental policy tomorrow. Environmental heroes "exhibit moral fortitude, acting with integrity in the face of pressure to do otherwise. Thoughtful and analytical, they bring wisdom to their actions and employ facts and the best available science to present their case . . . they exhibit courage in the face of seemingly insurmountable odds."[24] Let's continue forging ahead with what the nation's environmental heroes do best—making U.S. environmental policy by crafting solutions to address problems.

Notes

1. Tamara Thiessen, "How Clean Air Cities Could Outlast COVID-19 Lockdowns," *Forbes*, April 10, 2020, https://www.forbes.com/sites/tamarathiessen/2020/04/10/how-clean-air-cities-could-outlast-covid-19-lockdowns/#1f0d59 5b6bb5.

2. Helen Branswell, "A $30 Billion Gamble: Pandemic Expert Calls for Making COVID-19 Vaccines Before We Know They Work," *STAT*, April 7, 2020, https://www .statnews.com/2020/04/07/pandemic-expert-calls-for-making-coronavirus -vaccines-before-we-know-they-work/.

3. Sara Rinfret, Denise Scheberle, and Michelle Pautz, *Public Policy: A Concise Introduction* (Thousand Oaks, CA: CQ Press, 2018).

4. Michelle Pautz and Sara Rinfret, *The Lilliputians of Environmental Regulation* (New York: Routledge, 2013).

5. Sara Rinfret and Michelle Pautz, *Environmental Policy in Action* (New York: Palgrave, 2019).

6. Denise Scheberle, *Industrial Disasters and Environmental Policy* (New York, Routledge, 2018), 177.

7. Harold D. Lasswell, *Psychopathology and Politics* (Chicago: University of Chicago Press, 1902).

8. Sara Rinfret, "Cleaning Up the Air: The EPA and Shuttle Diplomacy," *Environmental Practice* 13, no. 3 (2011): 227–234, https://www.cambridge.org/core/journals/environmental-practice/article/cleaning-up-the-air-the-epa-and-shuttle-diplomacy/428626505DC3A3EC4759E07A4E3D0BF9.

9. Robert V. Percival, "Regulatory Evolution and the Future of Environmental Policy," *University of Chicago Legal Forum* 1997, no. 1, https://chicagounbound.uchicago.edu/cgi/viewcontent.cgi?article=1227&context=uclf.

10. Daniel J. Fiorino, *The New Environmental Regulation* (Cambridge, MA: MIT Press, 2006); Rinfret and Pautz, *Environmental Policy.*

11. Pautz and Rinfret, *Lilliputians of Environmental Regulation.*

12. Drew Calvert, "Navigating the Multigenerational Workplace," *Government Executive*, May 22, 2015, https://www.govexec.com/management/2015/05/navigating-multigenerational-workplace/113581/.

13. Sara Rinfret and Jeffrey Cook, "Trial by Fire? Rulewriters, Training, the BLM, and EPA," *Environmental Practice* 21, no. 3 (2019): 113–120, https://www.tandfonline.com/doi/abs/10.1080/14660466.2019.1656480.

14. National Park Service, "NPS Fundamentals," https://mylearning.nps.gov/program-areas/programs/nps-fundamentals/.

15. Marissa Martino Golden, "Interest Groups in the Rule-Making Process: Who Participates? Whose Voices Get Heard?" *Journal of Public Administration Research and Theory* 8, no. 2 (1998): 245–270, https://academic.oup.com/jpart/article-abstract/8/2/245/965781?redirectedFrom=fulltext.

16. Sara Rinfret, "Cleaning Up the Air: The EPA and Shuttle Diplomacy," paper presented to the Western Political Science Association Annual Meeting, San Antonio, TX, February 2011, https://papers.ssrn.com/sol3/papers.cfm?abstract_id=1766676.

17. Cameron Holley and Darren Sinclair, "Regulation, Technology, and Water: 'Buy-In' as a Precondition for Effective Real-Time Advanced Monitoring, Compliance and Enforcement," *George Washington Journal of Energy and Environmental Law* 7, no. 1 (2016): 52–66.

18. David Hindin, "Using Next Generation Compliance Drivers in Permits and Rules," paper presented at the J. B. and Maurice C. Shapiro Environmental Law Symposium on Advanced Monitoring, Remote Sensing, and Data Gathering, Analysis and Disclosure in Compliance and Enforcement, Washington, DC, March 2015.

19. Stephanie Kanowitz, "Environmental Inspectors Find Mobile App to Be a Force Multiplier," *GCN*, March 15, 2018, https://gcn.com/articles/2018/03/15/mobile-inspection-app.aspx.

20. David Morgan, Paul Lienert, and Timothy Gardner, "How Regulators Caught Volkswagen's Diesel Deception," *Insurance Journal*, September 25, 2015, https://www.insurancejournal.com/news/national/2015/09/25/382762.htm.

21. Regulations.gov, *Introducing Regulations.gov Beta*, June 12, 2019, video, 2:38, https://www.youtube.com/watch?v=WUq53-0PzMk.

22. Lisa Friedman, "E.P.A., Citing Coronavirus, Drastically Relaxes Rules for Polluters," *New York Times*, March 26, 2020, https://www.nytimes.com/2020/03/26/climate/epa-coronavirus-pollution-rules.html.

23. Liat Olenick and Alessandro Dal Bon, "Why You Should Care about Earth Day during the Coronavirus Pandemic," *Teen Vogue*, April 20, 2020, https://www.teenvogue.com/story/why-care-earth-day-during-coronavirus?fbclid=IwAR2V8dKuGGnvanekDxGDq9KMwGuQI6PfSKE9R4XynHibpzTmgCYV-_fZx50.

24. Scheberle, *Industrial Disasters.*

Suggested Readings

Kendrick, Rob. "A Running List of How President Trump Is Changing Environmental Policy." *National Geographic*, May 3, 2019. Available at https://www.nationalgeographic.com/news/2017/03/how-trump-is-changing-science-environment/.

Stokes, Leah. *Short Circuiting Policy.* Oxford: Oxford University Press, 2020.

Tesley, Albert. *The ABCs of Environmental Regulation.* 3rd ed. Lanham, MD: Bernan Press, 2016.

Bibliography

Branswell, Helen. "A $30 Billion Gamble: Pandemic Expert Calls for Making COVID-19 Vaccines before We Know They Work." *STAT*, April 7, 2020. Available at https://www.statnews.com/2020/04/07/pandemic-expert-calls-for-making-coronavirus-vaccines-before-we-know-they-work/.

Calvert, Drew. "Navigating the Multigenerational Workplace." *Government Executive*, May 22, 2015. Available at https://www.govexec.com/management/2015/05/navigating-multigenerational-workplace/113581/.

Finley, Bruce. "Colorado Sees 'Significant Declines' in Air Pollution as Coronavirus Ramps Down Driving, Industrial Activity." *Greeley Tribune*, April 5, 2020. Available at https://www.greeleytribune.com/news/colorado-sees-significant-declines-in-air-pollution-as-coronavirus-ramps-down-driving-industrial-activity/

Fiorino, Daniel J. *The New Environmental Regulation.* Cambridge, MA: MIT Press, 2006.

Friedman, Lisa. "E.P.A., Citing Coronavirus, Drastically Relaxes Rules for Polluters." *New York Times*, March 26, 2020. Available at https://www.nytimes.com/2020/03/26/climate/epa-coronavirus-pollution-rules.html.

Golden, Marissa Martino. "Interest Groups in the Rule-Making Process: Who Participates? Whose Voices Get Heard?" *Journal of Public Administration Research and Theory* 8, no. 2 (1998): 245–270. Available at https://academic .oup.com/jpart/article-abstract/8/2/245/965781?redirectedFrom=fulltext.

Hindin, David. "Using Next Generation Compliance Drivers in Permits and Rules." Paper presented at the J. B. and Maurice C. Shapiro Environmental Law Symposium on Advanced Monitoring, Remote Sensing, and Data Gathering, Analysis and Disclosure in Compliance and Enforcement, Washington, DC, March 2015.

Holley, Cameron, and Darren Sinclair. "Regulation, Technology, and Water: 'Buy-In' as a Precondition for Effective Real-Time Advanced Monitoring, Compliance and Enforcement." *George Washington Journal of Energy and Environmental Law* 7, no. 1 (2016): 52–66.

Kanowitz, Stephanie. "Environmental Inspectors Find Mobile App to Be a Force Multiplier." *GCN*, March 15, 2018. Available at https://gcn.com/articles/2018 /03/15/mobile-inspection-app.aspx.

Lasswell, Harold D. *Psychopathology and Politics.* Chicago: University of Chicago Press, 1902.

National Park Service. "NPS Fundamentals." Available at https://mylearning.nps .gov/program-areas/programs/nps-fundamentals/.

Olenick, Liat, and Alessandro Dal Bon. "Why You Should Care about Earth Day during the Coronavirus Pandemic." *Teen Vogue*, April 20, 2020. Available at https://www.teenvogue.com/story/why-care-earth-day-during-coronavirus? fbclid=IwAR2V8dKuGGnvanekDxGDq9KMwGuQl6PfSKE9R4XynHibpz TmgCYV-_fZx50.

Pautz, Michelle, and Sara Rinfret. *The Lilliputians of Environmental Regulation.* New York: Routledge, 2013.

Percival, Robert V. "Regulatory Evolution and the Future of Environmental Policy." *University of Chicago Legal Forum* 1997, no. 1: 159–198. Available at https://chicagounbound.uchicago.edu/cgi/viewcontent.cgi?article=1227& context=uclf.

Regulations.gov. *Introducing Regulations.gov Beta.* June 12, 2019. Video, 2:38. Available at https://www.youtube.com/watch?v=WUq53-0PzMk.

Rinfret, Sara. "Cleaning Up the Air: The EPA and Shuttle Diplomacy." *Environmental Practice* 13, no. 3 (2011): 227–234. Available at https://www.cambridge .org/core/journals/environmental-practice/article/cleaning-up-the-air-the-epa -and-shuttle-diplomacy/428626505DC3A3EC4759E07A4E3D0BF9.

———. "Cleaning Up the Air: The EPA and Shuttle Diplomacy." Paper presented at the Western Political Science Association Annual Meeting, San Antonio, TX, February 2011. Available at https://papers.ssrn.com/sol3 /papers.cfm?abstract_id=1766676.

Rinfret, Sara, and Jeffrey Cook. "Trial by Fire? Rulewriters, Training, the BLM, and EPA." *Environmental Practice* 21, no. 3 (2019): 113–120. Available at https://www.tandfonline.com/doi/abs/10.1080/14660466.2019.1656480.

Rinfret, Sara, and Michelle Pautz. *Environmental Policy in Action*. New York: Palgrave, 2019.

Rinfret, Sara, Denise Scheberle, and Michelle Pautz. *Public Policy: A Concise Introduction*. Thousand Oaks, CA: CQ Press, 2018.

Savoie, Michael. "The State of Telepresence: Healthcare and Telemedicine." *Robohub*, November 20, 2015. Available at https://robohub.org/the-state-of-telepresence-healthcare-and-telemedicine/.

Scheberle, Denise. *Industrial Disasters and Environmental Policy*. New York: Routledge, 2018.

St Anne, Michelle. "Confronting Crises: Corona and Climate Series." University of Sydney, Sydney Environment Institute, April 8, 2020. Available at http://sydney.edu.au/environment-institute/news/confronting-crises-corona-and-climate-series/?fbclid=IwAR0aVaIM5b3ezw6Ndy6t6_os0-5RSdlPW5I11JLCLghVV98lSiCVy3evrbk.

Thiessen, Tamara. "How Clean Air Cities Could Outlast COVID-19 Lockdowns." *Forbes*, April 10, 2020. Available at https://www.forbes.com/sites/tamarathiessen/2020/04/10/how-clean-air-cities-could-outlast-covid-19-lockdowns/#1f0d595b6bb5.

About the Contributors

Jeffrey J. Cook is a policy analyst at the National Renewable Energy Laboratory in Golden, Colorado. He received his Ph.D. in political science from Colorado State University in 2017, where he was an instructor of environmental and public policy courses. He has published articles relating to environmental and energy regulation in journals such as *Environmental Politics, Review of Policy Research*, and *Public Administration Quarterly*.

Deserai A. Crow is an associate professor in the School of Public Affairs at the University of Colorado Denver. She researches local and state-level environmental policy, including stakeholder participation and influence, information sources used, and policy outcomes. Her work focuses on environmental regulation and natural disaster recovery and risk mitigation in local communities and natural resource agencies. Her disaster research has been funded by the National Science Foundation. Crow earned her Ph.D. in environmental policy from Duke University's Nicholas School of the Environment. She also holds a master of public administration degree from the University of Colorado Denver's School of Public Affairs and a B.S. in journalism from the University of Colorado Boulder.

Charles Davis is a professor emeritus of political science at Colorado State University. His teaching and research interests lie within U.S. energy and environmental policy issues. His publications include books dealing with public lands policies and hazardous waste politics as well as book chapters and articles appearing in journals such as *Policy Sciences, Review of Policy Research, Society and Natural Resources, Polity, Administration and Society, Environmental Science and Technology,* and the *Journal of Forestry.* He received a Ph.D. from the University of Houston and a B.S. from Colorado State University.

Robert J. Duffy is a professor of political science at Colorado State University. His research focuses on environmental and energy issues as well as the role of organized interests in federal elections. His research has been published in *Environmental Politics, Administration and Society, Polity,* and the *Oxford Handbook of U.S. Environmental Policy.* He has authored or coauthored four books, including *Nuclear Politics in America: A History and Theory of Government Regulation, The Green Agenda in American Politics: New Strategies for the 21st Century, and Integrating Climate Energy and Air Pollution Policy,* with Gary Bryner. He holds a Ph.D. from Brandeis University, an M.A. from the University of Delaware, and a B.A. from Lafayette College.

Sara K. Guenther is a research scientist in the Department of Political Science at Montana State University. She studies executive politics, political communication, and public opinion, with substantive focus in environmental policy. Her dissertation examined the political circumstances contributing to the relaxation of federal standards under the Endangered Species Act. Before earning her Ph.D. in political science, Guenther studied marine mammal acoustics.

Lydia A. Lawhon is a lecturer with the Masters of the Environment Program at the University of Colorado Boulder, where she also earned her Ph.D. in environmental studies. Her research considers the role of local knowledge in crafting environmental policy as well as ways to reduce conflict and improve outcomes for controversial natural resources policy problems. Her dissertation focused on the transition to state management of wolves in Wyoming following federal delisting. Lawhon is also a research associate with the Northern Rockies Conservation Cooperative in Jackson, Wyoming. She holds a B.A. from Dartmouth College and a master's degree in environmental management from the Yale School of the Environment.

Michelle C. Pautz is a professor of political science and assistant provost for the Common Academic Program at the University of Dayton. Her research largely focuses on two areas: the implementation of environmental regulation, particularly at the state level; and the portrayal of bureaucracy and bureaucrats in contemporary American cinema and its effects on audiences. She has authored more than two dozen articles in journals ranging from *Administration and Society, Policy Studies Journal,* and *Review of Policy Research* to *Public Voices, PS: Political Science and Politics,* and the *Journal of Political Science Education.* She has authored seven books, including *Civil Servants on the Silver Screen: Hollywood's Depiction of Government and Bureaucrats, The Lilliputians of Envi-*

ronmental Regulation: The Perspective of State Regulators (with Sara R. Rinfret), and *US Environmental Policy in Action* (now in its second edition, also with Sara R. Rinfret). In 2016 she was recognized with the University of Dayton's Alumni Award for Outstanding Teaching. She holds a Ph.D. in public administration and an M.P.A. from Virginia Tech and a B.A. in economics, political science, and public administration from Elon University.

Sara R. Rinfret is a professor and associate dean of Public Administration and Policy at the University of Montana's Alexander Blewett School of Law. Her research is focused on environmental regulations and the scholarship of teaching and learning. More specifically, she is interested in the interactions between agencies and interest groups during the stages of environmental rule-making at the federal and state level, and the role of women in regulatory policy. Her work has been published in several journals, and she has coauthored four books: *The Lilliputians of Environmental Regulation: The Perspective of State Regulators, U.S. Environmental Policy in Action* (with Michelle Pautz), *Public Policy: A Concise Introduction* (with Michelle Pautz and Denise Scheberle), and the *Environmental Case* (with Judith Layzer). She was a Fulbright Specialist Program in public administration and studied with scholars at the University of Aarhus (Denmark) in 2016. In 2018, she was selected by University of Montana alumni as the "most inspirational teacher" of the year. She holds a Ph.D. from Northern Arizona University, an M.P.A. from the Ohio State University's John Glenn College, and a B.A. from Otterbein College (University).

Index

Page numbers followed by the letter t refer to tables. Page numbers followed by the letter f refer to figures.

www.ingramcontent.com/pod-product-compliance
Lightning Source LLC
Chambersburg PA
CBHW022357280326
41935CB00007B/215